Ethics and Integrity in Educational Contexts

Volume 8

The aim of this series is to provide an authoritative series of books on topics relating to ethics and integrity in educational contexts. Its scope includes ethics and integrity, defined in broad and inclusive terms, in educational contexts. It focuses on higher education, but also welcomes contributions that address ethics and integrity in primary and secondary education, non-formal educational contexts, professional education, etc. We welcome books that address traditional academic integrity topics such as plagiarism, exam cheating, and collusion.

In addition, we are particularly interested in topics that extend beyond questions of student conduct, such as

- Quality assurance in education;
- Research ethics and integrity;
- Admissions fraud;
- Fake and fraudulent credentials;
- Publication ethics;
- Educational technology ethics (e.g., surveillance tech, machine learning, and artificial intelligence, as they are used in education);
- Biomedical ethics in educational contexts;
- Ethics in varsity and school sports.

This series extends beyond traditional and narrow concepts of academic integrity to broader interpretations of applied ethics in education, including corruption and ethical questions relating to instruction, assessment, and educational leadership. It also seeks to promote social justice, diversity, equity, and inclusion.

The series provides a forum to address emerging, urgent, and even provocative topics related to ethics and integrity at all levels of education, from a variety of disciplinary and geographical perspectives.

Lisa Vogt • Brenda M. Stoesz
Editors

Academic Integrity in Vocational and Polytechnic Education

Springer

Editors
Lisa Vogt
Civil Engineering Technology
Red River College Polytechnic
Winnipeg, MB, Canada

Brenda M. Stoesz
Centre for the Advancement of Teaching and Learning
University of Manitoba
Winnipeg, MB, Canada

ISSN 2731-779X ISSN 2731-7803 (electronic)
Ethics and Integrity in Educational Contexts

ISBN 978-3-031-90090-7 ISBN 978-3-031-90091-4 (eBook)
https://doi.org/10.1007/978-3-031-90091-4

This Springer imprint is published by the registered company Springer Nature Switzerland AG
The registered company address is: Gewerbestrasse 11, 6330 Cham, Switzerland

Foreword

This book brings exciting new insights to the field of academic integrity. To date, much of what has been written about academic integrity has neglected to include the wisdom and expertise of individuals working in vocational and polytechnic education. Academic integrity discourse has historically been dominated by those who work in universities. This book changes that by elevating and amplifying the voices of those who work in community colleges, polytechnics, and other organizations that offer important and valuable education to learners in trades, technical, creative, and caregiving professions.

It is essential to discuss academic integrity within vocational and polytechnic education to prepare students for the ethical expectations of their future careers and protect public safety. Although there may be an assumption that vocational and polytechnic institutions focus solely on practical skills, students can also encounter written assessments, online exams, and collaborative projects where academic misconduct can occur. The contributors to this book argue that integrating academic integrity into faculty training, developing context-specific policies, creating authentic workplace-relevant assessments, and adopting restorative approaches to misconduct are foundational to developing a culture of integrity in vocational and trades education. By connecting the dots between honesty, responsibility, and ethical decision-making, educators in vocational and polytechnic contexts can equip students with the necessary skills for professional success and positive societal contributions.

Academic integrity is culturally and contextually bound. This includes institutional culture and the context of specific or specialized kinds of education. The contributors to this book explore the multifaceted nature of academic integrity within vocational and polytechnic education, challenging the assumption that academic integrity is less relevant in this context compared to universities. They highlight the importance of moving beyond a punitive approach to academic misconduct and instead fostering a culture of integrity that permeates all levels of vocational and polytechnic institutions, from policies to individual interactions. The chapter authors emphasize the need for clear communication of expectations, authentic assessments that are relevant to the workplace, and support for students in developing their understanding of academic integrity. They also explore the impact of external factors, such as industry standards and technological advancements, on academic integrity in vocational and polytechnic education, calling for ongoing adaptation and collaboration among stakeholders to address emerging challenges.

Collectively, the contributors to this edited volume stress that we should care about academic integrity in vocational and polytechnic education because it is not just an abstract concept, but rather, that ethical decision-making has real-world implications for student success, public safety, and the reputation of vocational and polytechnic institutions.

A key message woven throughout this book is that technical proficiency without ethical grounding is insufficient and potentially dangerous. Imagine, for instance, a healthcare worker trained through a technical or vocational program who, having plagiarized their way through assessments, lacks the fundamental knowledge needed to administer medication safely. This is just one example showing why fostering a culture of integrity is paramount. Academic integrity is about more than simply catching those who cheat in school; it requires a proactive approach that includes embedding academic integrity into faculty training, crafting clear policies specific to the context, and designing authentic assessments that mirror real-world workplace scenarios.

The chapter authors also illustrate how the rise of sophisticated and rapidly developing technologies, especially generative artificial intelligence, demands a nuanced approach. For example, although architectural technology instructors can use software to identify if students submit identical work, educators in other fields, such as music production, need to find ways to teach copyright and ethical collaboration in the age of easily manipulated and shared digital files. Ultimately, the chapters provide a roadmap for vocational and polytechnic educators and institutional leaders to not only prevent academic misconduct, but to prepare graduates who understand that integrity is not just about following rules, but about making responsible choices that reflect the ethical standards of their chosen professions.

Lisa Vogt and Brenda M. Stoesz have provided leadership and vision to make this book a reality. The contributing authors have laid a foundation for important dialogue, pedagogical innovation, and policy improvement to ensure that ethical decision-making is paramount in vocational and polytechnic education. This book is the first of its kind in the world and we are so pleased to include it in the Springer series, Ethics and Integrity in Educational Contexts.

Werklund School of Education
University of Calgary,
Calgary, Canada

Sarah Elaine Eaton

Preface

The invitation to take on the editorship of a book about academic integrity in vocational and polytechnic education should not have taken Lisa Vogt by surprise, given that Dr. Sarah Elaine Eaton has a knack for challenging academic integrity community members to tackle new projects, but it did. The invitation arrived in Lisa's inbox on a cold Canadian winter day, and she immediately began to consider what this type of project would involve and how it could connect to her work. With academic publishing not a requirement for her faculty position at a polytechnic institution in central Canada, she presented the opportunity to her manager, who recognized the value of this opportunity and encouraged her to accept. Lisa got to work organizing a two-year timeline from start to publication, reading relevant literature, and recruiting and supporting chapter authors.

As the original and sole editor at the time, Lisa started by identifying ideas and potential authors who could shape the content, articulate key points, and highlight what made academic integrity in vocational and polytechnic education unique. Lisa brainstormed ideas with academic integrity community members and was introduced to more potential contributors. Lisa's goal was to include stories from a global context; however, connecting with staff and faculty outside of North America and Australia in the vocational and polytechnic education sectors was challenging.

In June 2023, the Canadian Symposium on Academic Integrity (CSAI) was held at the University of Manitoba, with Dr. Brenda M. Stoesz at the helm of the planning committee. The two-day in-person event was a marked success, with many academic integrity community members from around the world congregating in Winnipeg, Canada. Coming together in person was electrifying, and a welcome change from the online conferences and workshops held throughout the COVID-19 pandemic. Many attendees felt like old friends since we had spent Monday mornings together in the *Academic Integrity Hour in Canada* online community of practice. This energy fueled the final recruitment drive to round out the proposal for this book.

The proposal included seasoned academic writers, first-time authors, and those in between. With approval from Springer, Lisa and the recruited authors set to work, following a 9-month schedule of collaborative brainstorming sessions in small online groups, drafting, peer editing, and external review. The task was no small feat, and Lisa was grateful to have Brenda join as co-editor in reviewing chapters, providing feedback to authors, and demonstrating editorial skills. The community of authors worked to support each other through collaboration and a spirit of generosity, each believing that documenting academic integrity in the vocational and polytechnic education sectors was important. The result is 16 chapters detailing experiences ranging from institutional practices to personal interactions. A Canadian contingent held the stronghold position, authoring the bulk of the chapters, with one Australian joining the group. The goal to include a more global perspective remains a task for another project.

Although the prominence of Canadian voices reflects the authors' community of practice, Canada is the Organisation for Economic Co-operation and Development (OECD) member country with the highest proportion of adults whose highest level of education is from short-cycle tertiary education such as vocational and polytechnic institutions (Skolnik, 2021a). The geographic distances between Canadian towns and cities result in highly collaborative communities in post-secondary education (Skolnik, 2018). Post-secondary professionals attend conferences together, recruit students side by side, and share learning materials, making Canadians well-positioned to take the lead in this discussion.

Fine-Tuning the Language

The Sectors

One of the challenges was selecting the appropriate name to describe the sectors of post-secondary education that are typically completed in two years or less, result in certificates or diplomas, and often teach applied and industry-focused skills. In Canada, these types of institutions have been called institutes, community colleges, and colleges. More recently, there has been a move in Canada for these institutions to take on a polytechnic identity that focuses on collaborative relationships with industry partners and applied research (Komesch & Watts-Rynard, 2021). Students in these institutions typically do not complete general education courses but enter directly into their chosen field of study (e.g., welding, electrical engineering technology, information security, culinary arts). Because education is governed at the provincial level in Canada (as stated within the constitution), each province determines the name and function of such institutions, setting them apart by geographic region and local industry demands (Marshall, 2008). As the market shifts, many of these institutions have started to offer 4-year degree programs, postgraduate certificates and diplomas, and shorter micro-credentials, making it even more difficult to set clear parameters about what they are and what they are not (Trotter & Mitchell, 2018).

In the United States, however, community colleges and colleges may offer specific vocational programs and general education programs that lead to associate degrees or transfer to a four-year college (Boggs, 2012; Skolnik, 2021b). In Canada, only the provinces of British Columbia and Alberta established colleges with the intent to create transfer pathways between colleges and universities (Dennison & Gallagher, 1986). In other provinces and territories, transfer and articulation agreements between any type of post-secondary institution vary by institution, program of study, and established provincial or institutional agreements (Skolnik & Jones, 1993; Skolnik, 2011). The term *vocational education and training* or *VET* is used more commonly in other parts of the world, where students can often choose to enter VET during their high school years or join an apprenticeship pathway later on (Markowitsch & Hefler, 2019).

Examining this wide-ranging context made it clear that one word could not define the context of focus in this book, so Lisa chose two. Lisa selected "vocational" because of its connection to workplace-specific training and its longstanding relevance to many institutions participating in this project. She also selected "polytechnic" because of its focus on industry and applied research, acknowledging its existing and newly developing importance to many institutions participating in this project. Each author described their context to help situate the reader in their environment.

Academic Integrity

Vocabulary to define academic integrity and related concepts varies slightly in policy and practice, depending on the institution. Members of the institutions represented in this project will recognize the definition of academic integrity put forward by the International Center for Academic Integrity – "a commitment, even in the face of adversity, to six fundamental values: honesty, trust, fairness, respect, responsibility, and courage" (International Center for Academic Integrity [ICAI], 2021, p. 4) but may take a range of approaches to describe an incident that lacks academic integrity. Some common options are academic misconduct, academic integrity breach, and referring to the specific type of breach such as cheating, collusion, inappropriate collaboration, or contract cheating. The authors of this collective work have used the language appropriate to their contexts, so readers will notice that the terms are not consistently applied across chapters. What remains important is that academic integrity describes positive behaviors rather than describing *both* desirable and undesirable behaviors. Take note and be prepared to clarify what academic integrity means the next time you hear that a student was "caught for academic integrity."

Themes

To develop the book's narrative, Lisa worked with authors to discuss potential themes and topics within their areas of expertise. Some contributors were clearly connected to projects that warranted documenting, whereas others had to think critically about how to present their thoughts, ideas, and experiences. As the chapters began to take shape, overlapping themes became evident. Three significant themes stand out for their presence throughout this work:

- **Prepare for a journey.** Practitioners describe academic integrity as a journey, noting the time and effort required to move forward with initiatives in a vocational or polytechnic institution, with multiple structural features that can slow and enhance the development of an academic integrity culture.
- **Make it real.** When students are studying for specific professions, concepts of integrity are not abstract. One of the greatest advantages of promoting academic integrity in vocational and polytechnic education may be the opportunity to link academic integrity to industry expectations of integrity.
- **It begins with us.** Each of us can foster academic integrity every day. We set the tone by demonstrating desired values and attitudes, holding others accountable, and showing compassion, thereby applying the hands-on and applied nature foundational to vocational and polytechnic education.

Organization of the Book

The book is organized into three sections:

Building and Maintaining Institutional Cultures of Academic Integrity

This section details journeys at five institutions to build and maintain institutional cultures of academic integrity. Munn et al. (2025) described a cultural shift spanning 10 years and involving a systems approach to making change in a small college environment. Ratcliffe et al. (2025) approached change to enhance institutional academic integrity from within the library at a mid-sized polytechnic. Meanwhile, Koren (2025) documented the work to shift perspectives on academic integrity in a large state-wide VET institution. To maintain cultures of academic integrity, Scurr (2025) provided insights into training via a micro-credential for faculty who are front-line decision-makers. Fukushima and Couteret (2025) rounded out this section by advocating for polytechnics to leverage the primary functions of polytechnic education (e.g., hands-on learning, industry-driven education, and innovation) in pursuing restorative practices that create a relational culture.

Impact of Structure and Organization of Vocational and Polytechnic Education on Academic Integrity

The second section explores how the structure and organization of vocational and polytechnic institutions have a specific impact on academic integrity. Ferguson and Toye (2025) examined the effect of stress through aspects inherent in a community college. Lucien-Bay et al. (2025) identified further pinch points observed by student conduct officers in a polytechnic. To investigate methods to reduce the burden of academic integrity and academic misconduct work in colleges and polytechnics, Schneider (2025) detailed the need for instructor support through compensation, administration, and professional development. Seeland et al. (2025) applied the 4M Framework to examine the structure and organization of nursing education and the nursing profession more broadly, setting an example for other industry-focused programs to follow. To conclude this section, Déri et al. (2025) provided insights into the connections between the fundamental values of academic integrity (ICAI, 2021) and the Canadian Military Ethos, again demonstrating the importance of applying academic integrity to the language and expectations specific to an organization or industry.

Educators Support of Academic Integrity: Vocational Training Vignettes

The third section documents the voices of leaders, educators, and support staff who support academic integrity in specific industry-focused environments. Butterworth (2025) shared strategies for engaging students in geomatics engineering technology by offering choices and encouraging creativity. MacIssac (2025) conducted a study investigating how instructors support collaborative learning while avoiding collusion in trade shop classrooms. Fairbairn and Tilley (2025) detailed their responses to students creating, producing, and engineering digital audio when ownership and rights management have students thinking about industry integrity from the start. Urbanski and Vogt (2025) discussed the power of exploratory conversations in a technical environment where computer-aided design (CAD) software can clearly show evidence of academic misconduct. Jackson (2025) reflected on how demonstrating and acknowledging the fundamental values of academic integrity (ICAI, 2021) in language-focused workshops for hospitality students can enhance professional practice and contribute to student success. Finally, Gray and Vogt (2025) applied a duty of care to create a compassionate and empathetic culture in automotive technologies.

Education Brings Challenges

The collaboration between Lisa, from the polytechnic sector, and Brenda, from the university sector revealed holes in the attempt to differentiate vocational and polytechnic education from university education. Both contexts operate labs, practicums, and other hands-on environments, and the challenges are not entirely unique either. Students enter an educational environment to learn, no matter where that is or the knowledge and skills they hope to acquire. Instructors will confront challenges in their roles as educators, no matter what they are teaching. Rather than trying to differentiate vocational and polytechnic education from universities and other contexts, authors were advised to focus on clearly describing their context. In doing so, the similarities across education sectors became evident, reinforcing the idea that academic integrity is a conversation for everyone in education. As a result, this project concludes by confidently asserting that each education sector cannot operate in isolation from others. Collaboration across all areas of post-secondary education is essential, and there is much to gain from doing so.

Why, then, is it necessary to publish a book that focuses on vocational and polytechnic education? The phrase "nothing about us, without us," originally rooted in rights movements within marginalized communities (Charlton, 1998; Marsden et al., 2020), provides a guiding principle here. It underscores the importance of bringing unheard voices into the conversation and including the community members directly involved and affected. This book provides dedicated space for members of the vocational and polytechnic sectors to be active participants in developing best practices for academic integrity that are responsive to their environments. In turn, vocational and polytechnic education leaders, faculty, and staff can see their institutions, students, and experiences represented, creating more inclusive conversations. The result empowers members of vocational and polytechnic education communities to build on their strengths and develop more robust cultures of academic integrity—a topic explored in depth in Chapter 1.

The editorial team is thrilled to introduce *Academic Integrity in Vocational and Polytechnic Education*– a sincere invitation for professionals in these sectors to engage in meaningful, ongoing conversations about academic integrity.

Winnipeg, Canada

Lisa Vogt
Brenda M. Stoesz

References

Boggs, G.R. (2012). The evolution of the community college in America: Democracy's colleges. *Community College Journal, 82*(4). Pp. 36-39. https://www.ccjournal-digital.com/ccjournal/library/page/20120203/36/

Butterworth, C. (2025). Student engagement in geomatics engineering technology assessments. In L. Vogt and B. M. Stoesz (Eds.). Academic Integrity in Vocational and Polytechnic Education. Springer.

Charlton, J. I. (2004). *Nothing about us without us: Disability oppression and empowerment* (3rd Edition). University of California Press.
Dennison, J. D., & Gallagher, P. (1986). *Canada's community colleges: A critical analysis*. UBC Press.
Déri, C. E., Kirouac, C., & Moussu, L. (2025). Strength of character: Upholding academic integrity in Canadian military vocational training. In L. Vogt and B. M. Stoesz (Eds.). *Academic Integrity in Vocational and Polytechnic Education*. Springer.
Fairbairn, H., & Tilley, J. (2025). Automation, creativity, and ethics: Re-framing academic integrity in the training of music technology professionals. In L. Vogt and B. M. Stoesz (Eds.). Academic Integrity in Vocational and Polytechnic Education. Springer.
Ferguson, C. D., & Toye, M. A. (2025). Student stress in community colleges: Creating healthy, ethical spaces for learning. In L. Vogt and B. M. Stoesz (Eds.). Academic Integrity in Vocational and Polytechnic Education. Springer.
Fukushima, A., & Couteret, L. (2025). Building relational culture in polytechnic education through restorative approaches to academic integrity. In L. Vogt and B. M. Stoesz (Eds.). Academic Integrity in Vocational and Polytechnic Education. Springer.
Gray, C., & Vogt, L. (2025). A duty of care in automotive technology education. In L. Vogt and B. M. Stoesz (Eds.). Academic Integrity in Vocational and Polytechnic Education. Springer.
International Center for Academic Integrity (ICAI). (2021). *The fundamental values of academic integrity*. https://academicintegrity.org/images/pdfs/20019_ICAI-Fundamental-Values_R12.pdf
Jackson, E. (2025). English language support served with integrity: Reflections from a hospitality and tourism management workshop facilitator. In L. Vogt and B. M. Stoesz (Eds.). Academic Integrity in Vocational and Polytechnic Education. Springer.
Komesch, D., & Watts-Rynard, S. (2021). Workforce transformation: How polytechnics are positioning Canada for an uncertain future. In Polytechnic education: A vision for Ontario. Humber Press. https://pressbooks.pub/visionforontario/chapter/workforce-transformation-how-polytechnics-are-positioning-canada-for-an-uncertain-future/
Koren, N. (2025). Shifting perspectives: Academic integrity in vocational education and training (VET) in the age of generative artificial intelligence. In L. Vogt and B. M. Stoesz (Eds.). Academic Integrity in Vocational and Polytechnic Education. Springer.
Lucien-Bay, L., Couteret, L., & Radke, S. (2025). Polytechnic pinch points for academic integrity: Reflections from student conduct officers. In L. Vogt and B. M. Stoesz (Eds.). Academic Integrity in Vocational and Polytechnic Education. Springer.
MacIsaac, E. (2025). Exploring collusion as a form of academic misconduct in vocational shop classrooms. In L. Vogt and B. M. Stoesz (Eds.). Academic Integrity in Vocational and Polytechnic Education. Springer.

Markowitsch, J., & Hefler, G. (2019). *Future developments in vocational education and training in Europe: Report on reskilling and upskilling through formal and vocational education training.* Seville: European Commission, 2019, JRC 117691.

Marsden, N., Star, L., & Smylie, J. (2020). Nothing about us without us in writing: Aligning the editorial policies of the Canadian Journal of Public Health with the inherent rights of Indigenous Peoples. *Canadian Journal of Public Health, 111*(6), 822–825. https://doi.org/10.17269/s41997-020-00452-w

Marshall, D. (2008). Differentiation by degrees: System design and the changing undergraduate environment in Canada. *Canadian Journal of Higher Education, 38*(3), Article 3. https://doi.org/10.47678/cjhe.v38i3.499

Munn, C., Rexe, D., & Seeland, J. (2025). Successful institutional change: Lessons on building a culture of academic integrity. In L. Vogt and B. M. Stoesz (Eds.). Academic Integrity in Vocational and Polytechnic Education. Springer.

Ratcliffe, C., Macnaughton, S., Vuori, T., & Phillips, Y. (2025). Building an academic integrity plan: Our journey at a mid-sized polytechnic. In L. Vogt and B. M. Stoesz (Eds.). Academic Integrity in Vocational and Polytechnic Education. Springer.

Schneider, B. (2025). Advocating for fair compensation and employment equity: Reducing the burden of academic integrity and misconduct work in colleges and polytechnics In L. Vogt and B. M. Stoesz (Eds.). Academic Integrity in Vocational and Polytechnic Education. Springer.

Scurr, C. (2025). Faculty education on academic integrity: A micro-credential approach. In L. Vogt and B. M. Stoesz (Eds.). Academic Integrity in Vocational and Polytechnic Education. Springer.

Seeland, J., Cliplef, L., Hengen, M., Joyce, K., & Murray, K. (2025). From mega to micro: Collaborative strategies for enhancing academic integrity in a college nursing program. In L. Vogt and B. M. Stoesz (Eds.). Academic Integrity in Vocational and Polytechnic Education. Springer.

Skolnik, M. L. (2011). Re-conceptualizing the relationship between community colleges and universities using a conceptual framework drawn from the study of jurisdictional conflict between professions. *Community College Review, 39*(4), 352–375. doi:10.1177/0091552111424205

Skolnik, M.L. (2018). Dimensions of collaboration between community colleges and universities in facilitating attainment of baccalaureate degrees in career and technical education. In R. Latiner Raby & E. J. Valeau (Eds.), *Handbook of Comparative Studies on Community Colleges and Global Counterparts* (pp. 3–6). Springer International Publishing. doi:10.1007/978-3-319-50911-2_1

Skolnik, M. L. (2021a). Canada's high rate of short-cycle tertiary education attainment: A reflection of the role of its community colleges in vocational education and training. *Journal of Vocational Education & Training, 73*(4), 543–565. doi:10.1080/13636820.2020.1744692

Skolnik, M.L. (2021b). Issues in cross-national comparisons of institutions that provide vocational education and training. *Journal of Vocational Education & Training, 76*(1), pp. 87–105. doi:10.1080/13636820.2021.2008473

Skolnik, M. L., & Jones, G. A. (1993). Arrangements for coordination between university and college sectors in Canadian provinces. *Canadian Journal of Higher Education, 23*(1). https://doi.org/10.47678/cjhe.v23i1.183151
Trotter, L. D., & Mitchell, A. (2018). Academic drift in Canadian institutions of higher education: Research mandates, strategy, and culture. *Canadian Journal of Higher Education*, *48*(2), 92–108. https://doi.org/10.47678/cjhe.v48i2.188092
Urbanski, C., & Vogt, L. (2025). When technology uncovers the truth: Exploratory meetings in architectural technology. In L. Vogt and B. M. Stoesz (Eds.). Academic Integrity in Vocational and Polytechnic Education. Springer.

Acknowledgments

To the many supportive people who contributed to making this project possible:

Thank you to my family, Chris, Jonas, Mathew, and Livvy, for their patience and enthusiasm.

Thank you to Dr. Sarah Elaine Eaton and Dr. Brenda M. Stoesz, mentors who have continually believed in me and taught me so much.

Thank you to the leaders at RRC Polytech who believed in this endeavor and let me devote time to making it a reality – Lauren Phillips, Kerri Korabelnikov, Rebecca Molly Hiebert, and Jayne Geisel.

Thank you to each contributor for dedicating their time and energy to document their experience and practice. Extra kudos for supporting one another through ongoing discussions and peer editing.

Thank you to the additional peer editors who offered their insight and guidance to the chapter authors: Sheryl Boisvert, Jayne Geisel, Sarah Hie, Rebecca Molly Hiebert, Kerry Macdonald, Tasha Maddison, Lauren Phillips, Rick Robinson, Abhay Sharma, Nora Sobel, Rosemary Vogt, and Brea Williams.

With you, many things are possible.

Lisa Vogt

Red River College Polytechnic

Winnipeg, Manitoba, Canada

Over the past six years I have enjoyed working with Lisa in various capacities, including writing articles and giving presentations about ways to promote academic integrity, conducting research about students' knowledge and experiences related to academic integrity, and now this book. Lisa has accomplished so much in relatively little time as an expert in the field of academic integrity, and I am so proud to call her friend, colleague, collaborator, and now co-editor. Thank you, Lisa, for inviting me to join you on the last leg of this journey – an important collection of reflections, experiences, and expertise. Congratulations to all the authors for your perseverance and willingness to tell your vocational and polytechnic education stories! A welcome and much-needed addition to the academic integrity literature!

Brenda M. Stoesz

Centre for the Advancement of Teaching and Learning, University of ManitobaWinnipeg, Manitoba, Canada

We acknowledge that some of the chapter authors used generative artificial intelligence (genAI) tools to generate ideas, correct grammar, and evaluate the flow of ideas.

Contents

Editors and Contributors

About the Editors

Lisa Vogt is an instructor teaching ethics, leadership, and communication-focused courses in the Civil Engineering Technology department at Red River College Polytechnic in Winnipeg, Manitoba, Canada. She has dedicated over 20 years to higher education, beginning her career as an English as an Additional Language (EAL) teacher in Tokyo, Japan, before transitioning to Canadian higher education. Lisa has published and presented on academic integrity at local, national, and international venues, sharing research findings and practitioner insights. The common thread in her work is creating positive and inclusive educational environments that enable students and staff to thrive. She holds a B.A. in Intercultural Studies, a certificate in Teaching English to Speakers of Other Languages (TESOL), and an M.A. in Higher Education Administration and Leadership.

Brenda M. Stoesz is the Research Lead–Science of Teaching and Learning at the Centre for the Advancement of Teaching and Learning at the University of Manitoba in Winnipeg, Manitoba, Canada. Brenda began her teaching journey in 2000, teaching high school general science and biology before moving on to teach human anatomy and physiology at Assiniboine College's Licensed Practical Nursing program. Years later, during her graduate studies in psychology, her interest in teaching led to numerous teaching assistant, laboratory, and sessional instructor positions. As a researcher, Brenda's interests vary widely from academic integrity policy analysis to effectiveness of educational interventions, particularly how technologies (e.g., virtual reality, generative artificial intelligence) can be used to promote academic integrity, prevent academic misconduct, and enhance teaching and learning.

Contributors

Carina Butterworth Southern Alberta Institute of Technology (SAIT), Calgary, AB, Canada

Lynn Cliplef Assiniboine College, Brandon, MB, Canada

Linden Couteret University of Alberta, Edmonton, AB, Canada

Catherine E. Déri Université du Québec en Outaouais, Gatineau, QC, Canada

Hazel Fairbairn Douglas College, New Westminster, BC, Canada

Corrine D. Ferguson Bow Valley College, Calgary, AB, Canada

Alana Fukushima Northern Alberta Institute of Technology (NAIT), Edmonton, AB, Canada

Christine Gray Red River College Polytechnic, Winnipeg, MB, Canada

Mitchell Hengen Assiniboine College, Brandon, MB, Canada

Emilie Jackson Red River College Polytechnic, Winnipeg, MB, Canada

Kirk Joyce Assiniboine College, Brandon, MB, Canada

Caroline Kirouac Université du Québec à Rimouski, Lévis, QC, Canada

Nadia Koren TAFE NSW, St Leonards, NSW, Australia

Laura Lucien-Bay University of Alberta, Edmonton, AB, Canada

Emily MacIsaac Holland College, Charlottetown, PE, Canada

Sona Macnaughton Red Deer Polytechnic, Red Deer, AB, Canada

Lucie Moussu Royal Military College of Canada, Kingston, ON, Canada

Caitlin Munn Assiniboine College, Brandon, MB, Canada

Kealy Murray Assiniboine College, Portage la Prairie, MB, Canada

Yvonne Phillips Red Deer Polytechnic, Red Deer, AB, Canada

Susan Radke Southern Alberta Institute of Technology (SAIT), Calgary, AB, Canada

Caitlin Ratcliffe NorQuest College, Edmonton, AB, Canada

Deanna Rexe Assiniboine College, Brandon, Manitoba, Canada

Brooklin Schneider NorQuest College, Edmonton, AB, Canada

Cory Scurr Conestoga College, Kitchener, ON, Canada

Josh Seeland Assiniboine College, Brandon, MB, Canada

Brenda M. Stoesz The Centre for the Advancement of Teaching and Learning, University of Manitoba, Winnipeg, MB, Canada

Janette Tilley Douglas College, New Westminster, BC, Canada

Margaret A. Toye Bow Valley College, Calgary, AB, Canada

Chris Urbanski Civil Engineering Technology, Red River College Polytechnic, Winnipeg, MB, Canada

Lisa Vogt Civil Engineering Technology, Red River College Polytechnic, Winnipeg, MB, Canada

Teneil Vuori Red Deer Polytechnic, Red Deer, AB, Canada

Chapter 1
Building on Strengths: Nurturing Cultures of Academic Integrity in Vocational and Polytechnic Education

Lisa Vogt and Brenda M. Stoesz

Abstract Long-standing perceptions that the educational opportunities offered by vocational and polytechnic institutions are second-tier and inferior to the education provided by universities can have a significant impact on the decisions that students make about furthering their studies. Deficit thinking within and about vocational and polytechnic education also undermines the knowledge and skills of graduates of applied learning programs. Similarly, the foundation of the field of academic integrity originated from perceptions that students' deficits in their characteristics are risk factors for cheating. Shifting away from deficit models in education creates room for positive and empowering views of academic integrity that enable integrity cultures to develop. In this first chapter of *Academic Integrity in Vocational and Polytechnic Education*, we explore the notion of deficit thinking in vocational and polytechnic education and how appreciative inquiry and strengths-based education frameworks can support the success of academic integrity initiatives. We have highlighted the chapters in this volume that illustrated the positive effects of the transition in thinking from what is lacking to what works well by focusing on abilities, knowledge, and capacities that support the development of cultures of academic integrity within vocational and polytechnic education. In doing so, the authors have contributed to the inclusion of vocational and polytechnic education sectors in the broader academic integrity conversation and academic literature.

Keywords Academic integrity · Appreciative inquiry · Deficit thinking · Strengths-based education · Vocational education · Polytechnic institutions

---

L. Vogt (✉)
Civil Engineering Technology, Red River College Polytechnic, Winnipeg, MB, Canada
e-mail: lvogt@rrc.ca

B. M. Stoesz
Centre for the Advancement of Teaching and Learning, University of Manitoba, Winnipeg, MB, Canada
e-mail: Brenda.Stoesz@umanitoba.ca

L. Vogt, B. M. Stoesz (eds.), *Academic Integrity in Vocational and Polytechnic Education*, Ethics and Integrity in Educational Contexts 8,
https://doi.org/10.1007/978-3-031-90091-4_1

Building on Strengths: Nurturing Cultures of Academic Integrity in Vocational and Polytechnic Education

Post-secondary education often brings to mind university campuses, large lecture halls, and dormitories of students writing term papers, despite the fact that the label refers to all educational institutions that provide learning opportunities for students after their secondary or high school years. Outside of the university system are sectors of post-secondary education known as *vocational* and *polytechnic*, also referred to as institutes, community colleges, colleges, and vocational education and training [VET], depending on the region, country, and continent. Educational programs that can usually be completed in less than 4 years, focus on workplace-specific training and applied research, and have strong connections to industry often distinguish vocational and polytechnics institutions from universities.

Because the educational programming of vocational and polytechnic institutions typically enables individuals to enter into various careers directly after completing their training, employers must be assured that the qualifications earned by graduating students are valid measures of their knowledge and skills. The transition from study and/or experiential learning to the workforce is where the intersection between academic and workplace integrity is most easily observed—both environments require that individuals adhere to sets of ethical principles and standards that guide their actions in particular contexts. Learning with integrity, therefore, is vital in all post-secondary institutions, including those providing vocational and polytechnic education, to ensure that students' successes within them are true measures of their learning and are transferrable to their future careers.

Developing and committing to nurturing cultures of integrity takes time (Morris & Carrol, 2016), effort, patience, and perseverance. Those working in vocational and polytechnic education are acutely aware of the importance of academic integrity in meeting students' learning goals, employment and career aspirations, and the needs of employers but may find that the academic literature from which to draw strategies to promote integrity is limited (Ferguson et al., 2023; Hamilton & Wolsky, 2022). Acknowledging that educators in the vocational and polytechnic sectors may lack the opportunities, time, resources, and procedural knowledge to engage in research and scholarship on academic integrity, this edited volume provided a platform for higher education professionals to share their experiences, practices, strategies, and knowledge about academic integrity and contribute to the growing higher education literature (Murdoch & House, 2024). Authors documented their practices and the practices of their institutions, identifying obstacles while also describing how building on existing strengths nurtures cultures of academic integrity.

This first chapter of *Academic Integrity in Vocational and Polytechnic Education* explores the notion of deficit thinking in vocational and polytechnic education and how its members can shift their mindsets toward more positive and productive ways of teaching, developing policies and procedures, and other strategies to ensure that students learn with integrity. Evidence of the transition from what is lacking to focusing on abilities, knowledge, and capacities is found in various chapters within this volume. Chapter authors have illustrated the positive effects of these approaches to promoting academic integrity and dealing with academic misconduct on students, educators, and administrators.

Deficit Thinking within Vocational and Polytechnic Education

The perception that university study is superior to vocational and polytechnic education has existed for decades (e.g., Abrassart & Wolter, 2020; Edmunds, 2017; Raby & Valeau, 2018; Sappa et al., 2019; Taylor, 2020) and the sectors face ongoing challenges in being recognized as an equivalent option (Ewart, 2024; Watts-Rynard, 2024). The under-valuing of the knowledge and skills of vocational and polytechnic program graduates may be felt when colloquial expressions are uttered (e.g., 'grease monkey' referring to automotive technicians). Use of derogatory language implies that the educational opportunities offered by vocational and polytechnic institutions are second-tier or last-resort options; this can have an enormous impact on the decisions that students make about furthering their studies (Weiss, 2021). Marketing campaigns have attempted to counteract the disparaging narrative by promoting the benefits of applied learning to convince guidance counsellors, parents, and prospective students that vocational and polytechnic education leads to meaningful, challenging, and gainful employment. Meanwhile, local governments, funders, and industry leaders recognize that educating the future workforce is crucial to keeping society safe and well in occupations ranging from aircraft maintenance, building trades, and engineering technologies to healthcare, medical technologies, and early childhood education, among other vocations.

Deficit thinking may also appear in the minds of vocational and polytechnic educators and in role comparisons to professor-rank faculty members at research-intensive universities. In the vocational and polytechnic education sectors, many faculty members' primary role is educator—to plan lessons, teach classes, grade assessments, support students, and participate in departmental activities (Dennison & Gallagher, 1986)—a crucially important responsibility. Collective agreements between institutions and faculty associations may also stipulate a full-time teaching load for teaching faculty, which leaves little to no time to engage in academic research and dissemination activities. Obtaining release time from teaching duties to engage in academic research activities (which often take several months or years to complete) can also be challenging (Edmunds, 2017; Rosenkrantz, 2013). As such, management may actively discourage research and scholarship activities to adhere to collective agreements and attempt to prevent heavier workloads and time spent unremunerated potentially leading to negative impacts on the health and well-being of teaching faculty. Moreover, when academic research and scholarship are not expected and incentivized, managers at vocational and polytechnic institutions hire individuals who have specialized knowledge and skills in particular vocations and knowledge and skills in particular vocations and might not prioritize formal training in research methods and analyses during the hiring process. Where vocational and polytechnic institutions offer degree programs, faculty members may be formally trained in research methodology and do engage in research and scholarship activities.

These points should not be automatically perceived as barriers or challenges to overcome, as doing so contributes to the deficit narrative of the vocational and polytechnic education sectors. Nevertheless, fewer studies about academic integrity conducted within and in partnership with vocational and polytechnic institutions are published. This may perpetuate the myths that the vocational and polytechnic sectors are fundamentally different from and lesser than the university sector, and perhaps that academic misconduct is not a pressing issue and does not have an impact on faculty. In her chapter, Schneider (2025) debunks this notion by describing the profound impact of academic misconduct on faculty in vocational and polytechnic education. Furthermore, Lucien-Bay et al. (2025) illustrate how faculty fear tarnishing their relationships with students and industry partners if they identify and report academic cheating, creating additional nuances to the academic integrity cultures in vocational and polytechnic education that are not often documented in the literature.

Moving Forward from Deficit- to Strengths-Based Thinking

Deficit thinking is not limited to the perceptions of vocational and polytechnic education. Education as a whole is a problem-based field, seen as gaps to be studied and addressed or "fixed." The deficit mindset is so pervasive within education that it may be invisible and is therefore difficult to change even when attention has been drawn to this fact (see Cooperrider & Srivastva, 1987; Davis & Museus, 2019). Similarly, the foundation of the field of academic integrity originated from perceptions that students' deficits in their characteristics are risk factors for cheating and that they must bear all the blame for poor choices that lead to allegations of academic misconduct. Reinforcing negative perceptions about students and using fear-based messages about the dire consequences of academic cheating can lead to adversarial relationships and amplified combative feelings. Instructors (and institutions) hold positions of power over students; as such, when instructors and administrators demand compliance and punish overly harshly, psychologically unsafe spaces are created, which obstruct meaningful learning (Bertram Gallant, 2018; Brooks, 2024; Lindstrom, 2022). Shifting away from deficit models in education creates room for positive and empowering views of academic integrity that have enabled integrity cultures to grow in number over the past 30 years (Bertram Gallant & Rettinger, 2022).

To provide students with a solid foundation for learning, an academic integrity culture must be established and nurtured continuously within educational and training settings. Encouraging students to engage in desirable academic behaviours requires whole-hearted efforts by all members of vocational and polytechnic communities to develop, revise, and enact educative policies and procedures for academic integrity and implement effective strategies for learning (Bretag et al., 2011) that incorporate the values of honesty, trust, respect, fairness, responsibility, and courage (International Center for Academic Integrity [ICAI], 2021). In cultures of academic integrity, values are discussed openly and alongside holding individuals accountable for integrity violations—both acts support the development of a shared understanding of expectations moving forward (Fukushima & Couteret, 2025). Students, educators, and administrators of the post-secondary community (and society) can contribute to shifting conversations away from strictly punitive and rule-based methods to deter academic misconduct towards strengths-based and educative approaches to promote integrity (Bertram Gallant, 2011; Morris & Carroll, 2011). Appreciative inquiry and strengths-based education are relevant frameworks to describe and guide this shift.

Appreciative Inquiry

Appreciative inquiry is a change management approach that views organizations as living systems that require curiosity, nurture, and affirmation for purposes of improvement and possibilities rather than as possessing deficits and problems that must be corrected and solved (Cooperrider & Srivastva, 1987; Cooperrider & Whitney, 2005). In appreciative inquiry, emphasizing what is *not* working must be resisted in favour of identifying, describing, and building upon what works (see Fiorentino, 2012). Building on what *is* working requires the recognition of the circumstances that give life to the organization and enable its people and systems to function optimally (Cooperrider & Whitney, 2005; Whitney & Trosten-Bloom, 2010). Assessing the "inner dialogue" (Cooperrider et al., 2008, p. 20) of an organization can also help to dismantle deficit thinking in favour of positive, hope-filled discourse and actions to achieve the desired future state (Cooperrider & Whitney, 2005).

The appreciative inquiry process was developed as a four-step cycle (Cooperrider & Whitney, 2005) with the addition of a fifth step to define the scope and purpose of the inquiry (see Kessler, 2013) (see Fig. 1.1). In the *define* phase, the identification of the core focus of the inquiry sets the stage for a clear vision of the future steps. Working from this vantage point, the *discovery* phase involves identifying system

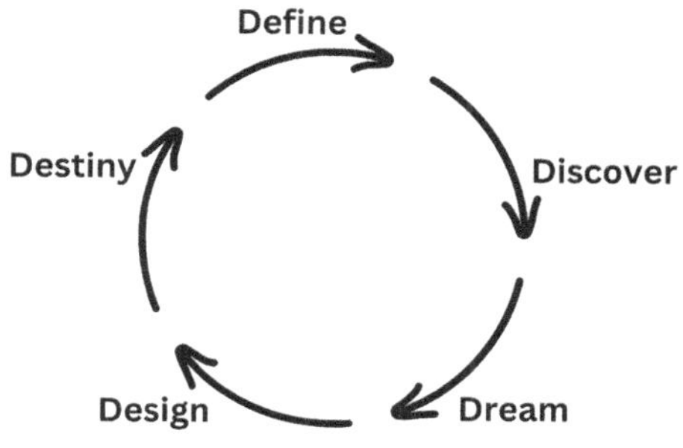

Fig. 1.1 5D Model of Appreciative Inquiry

components that already work well, highlighting exemplars and positive outcomes. Moving into the *dream* phase, hopes and aspirations are articulated to envision what an inspiring future could hold, which drives further possibilities. The *design* phase actions dreams by outlining concrete steps and strategies. Reaching the *destiny* phase signals the move toward implementing the vision, sustaining momentum, and evaluating outcomes. The cycle is grounded on positive future thinking with continuous learning, resulting in an iterative process that may repeat or move between stages to address current needs.

Verhoef et al. (2022) demonstrated how the community of practice (CoP) at North-West University in South Africa applied appreciative inquiry to ground their activities of discovery, sharing, and questioning to identify their individual and collective strengths, best practices for teaching and learning, and academic integrity. The CoP and its commitment to curiosity and exploration served to empower students and educators, create and nurture a culture of academic integrity, and was future-focused during a time of massive disruption (e.g., the COVID-19 pandemic). Advancements in various technologies over the last several decades have also challenged members of educational communities to think differently about teaching and learning. The most recent example is the release of accessible and free generative artificial intelligence tools, which continue to test our understanding of what it means to teach effectively, learn deeply, and assess accurately and has demanded adaptation of academic integrity initiatives (see Fairbairn & Tilley, 2025; Koren, 2025, in this volume). Fukushima and Couteret (2025), also in this volume, reason that the relationship building and repairing components of restorative practices are required to heal from the harms inflicted upon students and educators when academic integrity values are violated. Restorative practice requires collaboration, sharing, and relationship repair and aligns well with the positive future thinking cycles that are foundational to appreciative inquiry.

Strengths-Based Education

Similar to appreciative inquiry, strengths-based education focuses on what works, emphasizing positive efforts, achievements, and assets (Lopez & Louis, 2009) to foster well-being, deep learning, academic achievement, and healthy social development. Strengths-based education was built on the following principles:

- Measure strengths and positive characteristics
- Individualize approaches to respond to unique strengths
- Network with strengths at the forefront to bolster connections and relationships
- Deliberately apply approaches that will bring out strengths
- Intentionally develop strengths through new experiences, resources and opportunities

Strengths-based approaches have been applied in various ways to counter the deficit views of diverse student populations founded on Eurocentric ideals, instead leaning into characteristics and abilities (e.g., multilingualism, cultural knowledge, and social networks) that can enhance opportunities to succeed in studies and careers (Pang et al., 2018). Working from the perspective that individuals are capable and competent members who belong in community also supports the process of the internalization of values, whereby intrinsic motivation, rather than external pressures, drives behaviours that reflect adopted standards and values (Gagné & Deci, 2005).

Building on the five principles of strengths-based education, Eaton et al. (2024) proposed a sixth principle— synchronous and asynchronous communication—foundational to the work of the geographically distributed community of academic integrity researchers, educators, and higher education professionals in Canadian higher education (including universities and vocational and polytechnic education sectors). A large proportion of the academic integrity literature encourages educators, administrators, and other higher education professionals to recognize and build upon students' existing capabilities to guide them toward academic integrity expectations (see Bertram Gallant, 2017; Eaton, 2022; Howard, 1992), and the practices of the academic integrity community in Canada is grounded in its collective strengths.

Grounded in Strength and Curiosity

Humans are inclined to identify problems and solutions, making deficit thinking remarkably easy to fall into, which is observed in the emotionally taxing process of identifying, reporting, and resolving cases of alleged academic misconduct. Applying principles of appreciative inquiry and strengths-based education helps to reframe conversations and actions in positive directions. The chapter authors in this volume contribute to this literature by sharing their personal accounts and professional perspectives about what has worked well in their vocational and polytechnic institutions and how their educational communities are building on their successes. These stories will inspire readers to ask strengths-focused questions: *What already works well? Which voices and strategies should be amplified? What stories illustrate success?*

What Works Well?

Within any vocational and polytechnic institution (and higher education more broadly), members in various roles and with dual roles (e.g., administrators who teach) will feel discouraged and require time to process hardships and must receive support and be renumerated for their efforts (see Schneider, 2025, in this volume). The language that we use has the power to influence reality, and vocabulary that focuses on deficits blocks the chance for positive growth and vision in the future

(Cooperrider et al., 2008). This could be the first signal that a strengths-based approach is needed. Every context has strengths that can be identified and amplified, even aspects that appear to be deficits (e.g., student stress and mental health concerns) can be examined and perceived as an opportunity to identify and build upon *what is working well* to move toward learning spaces that are healthy, supportive, and ethical (see Ferguson & Toye, 2025, in this volume).

Chapters authored by Koren (2025), Munn et al. (2025), and Ratcliffe et al. (2025) highlight the academic integrity journeys of vocational and polytechnic institutions in Australia and Canada. The authors described different contexts, approaches, and timelines, resulting in multiple viewpoints on achieving similar goals. Scurr's (2025) account of developing a 5-part academic integrity micro-credential available at Conestoga College demonstrated that empowering faculty members through professional development opportunities is an effective strategy for fostering integrity culture. Learning about others' successes may spark new ideas for development or facilitate the identification of strengths, initiatives, strategies, and approaches that have already been shown effective in one's own context (Royer & Latz, 2016). In addition, integrating the successful approaches of others into existing collaborative programs or units (Jackson, 2025; Koren, 2025), cross-institutional partnerships (Munn et al., 2025; Ratcliffe et al., 2025), or CoPs (Schneider, 2025) can bolster their effectiveness for promoting academic integrity in teaching and learning contexts. Viewing identified opportunities and challenges through the lens of appreciative inquiry can make them personal and meaningful, and supports community development, and feelings of acceptance and belonging (Cooperrider & Whitney, 2005)—all crucial for ensuring further successes in fostering cultures of academic integrity.

Which Voices and Strategies Should Be Amplified and how?

After the identification of strengths and opportunities, the next steps involve leaning into them and generating buy-in from various stakeholders, including students. Buy-in can be generated by listening to educational community members describe their values and learning opportunities that are challenging and useful for obtaining employment or advancing careers (Herman & Loewenstein, 2017). Students, faculty, and administrators of vocational and polytechnic institutions can lean into the strength of connections between the specific learning and training program and the codes of ethics or regulatory bodies of the professions that students are training for and build on that connection by further aligning the learning and codes of ethics with the values of academic integrity (Déri et al., 2025; Seeland et al., 2025). The smaller cohorts of many courses and programs in this higher education sector is another important strength as it provides the foundation for the development of respect amongst and between students and instructors. Good relationships between educators and learners enhances learning and integrity. A key strength of vocational and polytechnic institutions is the direct partnership with industry. This partnership enables educators to tie the learning environment to workplace tasks and projects explicitly (MacIsaac, 2025) and can do so through individualized learning activities and assessments (Butterworth, 2025)—an academic integrity promoting strategy. In addition, by choosing industry partners who prioritize hiring graduates with well-developed personal philosophies consistent with workplace values (Fairbairn & Tilley, 2025; Fukushima & Couteret, 2025; Urbanski & Vogt, 2025), educators and administrators are more likely to be successful in getting buy-in from students. These are just a few examples of the strengths of vocational and polytechnic institutions and ways in which they can be capitalized on to promote learning and integrity.

What Stories Illustrate Success?

An ongoing focus on *what is lacking* often leads to missed opportunities to celebrate and amplify successes. Shifting away from negative patterns may take substantial time and effort and can be especially challenging in environments where people feel powerless. However, following the philosophy of "being the change you wish to see" (Lorrance, 1974) or "let it begin with me" (Al-Anon Family Groups, n.d.) means that each person can do something to highlight success. Appreciative inquiry interventions improve emotional well-being and self-confidence, ultimately strengthening academic professional identities (Miglianico et al., 2021). When you share your experiences and detail your passion and dedication for promoting academic integrity, you can offer immeasurable support (Gray & Vogt, 2025), adding much-needed value and recognition to the daily accomplishments of vocational and polytechnic education faculty (Sappa et al., 2019). This type of thinking rings true for strengths-based approaches and creates further opportunities to welcome the community to co-create hope for our future together (Royer & Latz, 2016).

Finding success stories may mean leaving your comfort zone and spanning boundaries to gather new ideas (MacIsaac, 2025). Educators and administrators could revisit long-standing practices with fresh eyes and attempt to identify *what is possible* by sharing their achievements with others (Jackson, 2025) and being open to receiving feedback to further strengthen initiatives, approaches, and strategies. The final stage of appreciative inquiry makes it clear that reaching the goal is not achieved overnight. Continued reflection and innovation are required to maintain and advance—an important reminder for growing a culture of academic integrity in vocational and polytechnic education.

The maturity of academic integrity culture can be measured in a variety of ways (see Bertram Gallant, 2022; Bertram Gallant & Drinan, 2008; Glendinning, 2014; Zhang et al., 2021), including whether the educational community has engaged in dissemination activities to share their successes with others. A plethora of publications describe academic integrity philosophy, education, and policy in universities, but academic integrity research and scholarship originating from the vocational and polytechnic education sectors has been rather limited. This can be problematic, especially when educators are unable to adapt and apply university strategies to their vocational or polytechnic teaching and learning contexts. As a result, they may reject effective strategies that are deemed unsuitable for colleges or may adopt the expectations of the university context; either action may not inappropriate.

Conclusion

Educators, higher education professionals, and researchers nurturing academic integrity in the vocational and polytechnic education sectors are actively breaking down assumptions, confronting deficit thinking, and developing strategies that align with the frameworks of appreciative inquiry and strengths-based education. This shift empowers members of these organizations and communities to create teaching and learning environments that are positive and inspiring and support those who work and learn within them, including students. Each contributor to this volume, *Academic Integrity in Vocational and Polytechnic Education*, invited readers to learn from their lived experiences, illuminating often unexamined aspects of post-secondary education. In doing so, they have contributed to the inclusion of vocational and polytechnic education in the broader academic integrity conversation and academic literature. By documenting the practices of the members of their educational communities, the authors not only showcased the diverse realities of these educational pathways but also provided an opportunity for others to see themselves reflected. Through the power of education and community, we can build a culture of integrity that we envision in the vocational and polytechnic education sectors and support the efforts of industry and society.

References

Abrassart, A., & Wolter, S. C. (2020). Investigating the image deficit of vocational education and training: Occupational prestige ranking depending on the educational requirements and the skill content of occupations. *Journal of European Social Policy, 30*(2), 225–240. https://doi.org/10.1177/0958928719855298

Al-Anon Family Groups. (n.d.). *The Al-Anon slogans*. https://al-anon.org/for-members/members-resources/literature/al-anon-slogans/#:~:text=%E2%80%9CAl-Anon%E2%80%99s%20slogan,%20%E2%80%98Listen%20and%20Learn%E2%80%99%20reminds%20us%20that%20if%20we

Bertram Gallant, T. (2011). *Building a culture of academic integrity* [White paper]. Magna Publications.. https://www.depts.ttu.edu/tlpdc/Resources/Academic_Integrity/files/academicintegrity-magnawhitepaper.pdf.

Bertram Gallant, T. (2017). Academic integrity as a teaching and learning issue: From theory to practice. *Theory Into Practice, 56*(2), 88–94. https://doi.org/10.1080/00405841.2017.1308173

Bertram Gallant, T. (2018). Part-time integrity? Contingent faculty and academic integrity. *New Directions for Community Colleges, 2018*(183), 45–54. https://doi.org/10.1002/cc.20316

Bertram Gallant, T. (2022, February 11). "When a thousand flowers bloom": The Academic Integrity Rating System (AIRS). *International Center for Academic Integrity Blog*. https://academicintegrity.org/resources/blog/96-2022/february-2022/345-when-a-thousand-flowers-bloom-the-academic-integrity-rating-system-airs?highlight=WyJhaXJzIl0=

Bertram Gallant, T., & Drinan, P. (2008). Toward a model of academic integrity institutionalization: Informing practice in postsecondary education. *Canadian Journal of Higher Education, 38*(2), 24–43. https://doi.org/10.47678/cjhe.v38i2.508

Bertram Gallant, T., & Rettinger, D. (2022). An introduction to 30 years of research on academic integrity. *Journal of College and Character, 23*(1), 1–5. https://doi.org/10.1080/2194587X.2021.2017975

Bretag, T., Mahmud, S., Wallace, M., Walker, R., James, C., Green, M., East, J., McGowan, U., & Patridge, L. (2011). Core elements of exemplary academic integrity policy in Australian higher education. *International Journal for Educational Integrity, 7*(2). https://doi.org/10.21913/IJEI.v7i2.759

Brooks, R. M. (2024). Integrity through experience: Fostering a culture of academic integrity through an experiential learning approach. In S. E. Eaton (Ed.), *Second handbook of academic integrity* (pp. 287–303). Springer Nature Switzerland. https://doi.org/10.1007/978-3-031-54144-5_94

Butterworth, C. (2025). Student engagement in geomatics engineering technology assessments. In L. Vogt & B. M. Stoesz (Eds.), *Academic integrity in vocational and polytechnic education*. Springer.

Cooperrider, D., & Srivastva, S. (1987). Appreciative inquiry in organizational life. *Research in Organizational Change and Development, 1*, 129–169.

Cooperrider, D. L., Stavros, J. M., & Whitney, D. (2008). *The appreciative inquiry handbook: For leaders of change* (2nd ed.). Berrett-Koehler Publishers.

Cooperrider, D. L., & Whitney, D. (2005). *Appreciative inquiry: A positive revolution in change*. Berrett-Koehler Publishers.

Davis, L. P., & Museus, S. D. (2019, July 19). *Identifying and disrupting deficit thinking*. Medium. https://medium.com/national-center-for-institutional-diversity/identifying-and-disrupting-deficit-thinking-cbc6da326995.

Dennison, J. D., & Gallagher, P. (1986). *Canada's community colleges: A critical analysis*. UBC Press.

Déri, C. E., Kirouac, C., & Moussu, L. (2025). Strength of character: Upholding academic integrity in Canadian military vocational training. In L. Vogt & B. M. Stoesz (Eds.), *Academic integrity in vocational and polytechnic education*. Springer.

Eaton, S. E. (2022, February 22). *Check your bias at the door.* University Affairs. https://www.universityaffairs.ca/features/feature-article/check-your-bias-at-the-door/.

Eaton, S. E., Stoesz, B. M., & McKenzie, A. (2024). Academic integrity leadership and community building in Canadian higher education. In S. E. Eaton (Ed.), *Handbook of academic integrity* (pp. 1–21). Springer International Handbooks of Education. Springer. https://doi.org/10.1007/978-3-031-54144-5_171

Edmunds, T. K. (2017). Perceived barriers to SME-college collaboration: The case of the province of Manitoba. *College Quarterly, 20*(2) http://collegequarterly.ca/2017-vol20-num02-spring/perceived-barriers-to-sme-college-collaboration-the-case-of-the-province-of-manitoba.html

Ewart, C. (2024, October 1). *Opinion: This is not the time to discourage students from choosing public colleges*. The Vancouver Sun. https://vancouversun.com/opinion/op-ed/opinion-this-is-not-the-time-to-discourage-students-from-choosing-public-colleges

Fairbairn, H., & Tilley, J. (2025). Automation, creativity, and ethics: Re-framing academic integrity in the training of music technology professionals. In L. Vogt & B. M. Stoesz (Eds.), *Academic integrity in vocational and polytechnic education*. Springer.

Ferguson, C. D., & Toye, M. A. (2025). Student stress in community colleges: Creating healthy, ethical spaces for learning. In L. Vogt & B. M. Stoesz (Eds.), *Academic integrity in vocational and polytechnic education*. Springer.

Ferguson, C. D., Toye, M. A., & Eaton, S. E. (2023). Contract cheating and student stress: Insights from a Canadian community college. *Journal of Academic Ethics, 21*, 685–717. https://doi.org/10.1007/s10805-023-09476-6

Fiorentino, L. H. (2012). Positive perspectives on the profession: Reframing through appreciative inquiry. *Quest, 64*(4), 209–228. https://doi.org/10.1080/00336297.2012.723464
Fukushima, A., & Couteret, L. (2025). Building relational culture in polytechnic education through restorative approaches to academic integrity. In L. Vogt & B. M. Stoesz (Eds.), *Academic integrity in vocational and polytechnic education*. Springer.
Gagné, M., & Deci, E. L. (2005). Self-determination theory and work motivation. *Journal of Organizational Behavior, 26*(4), 331–362. https://doi.org/10.1002/job.322
Glendinning, I. (2014, June 16–18). *Assessing maturity of institutional policies for underpinning academic integrity*. International Integrity and Plagiarism Conference, Coventry University. https://marketing-porg-statamic-assets-us-west-2.s3-us-west-2.amazonaws.com/main/Glendinning_paper%20(1).pdf
Gray, C., & Vogt, L. (2025). A duty of care in automotive technology education. In L. Vogt & B. M. Stoesz (Eds.), *Academic integrity in vocational and polytechnic education*. Springer.
Hamilton, M., & Wolsky, K. (2022). The barriers to faculty reporting incidences of academic misconduct at community colleges. In S. E. Eaton & J. Christensen Hughes (Eds.), *Academic integrity in Canada: An enduring and essential challenge* (pp. 467–485). Springer International Publishing. https://doi.org/10.1007/978-3-030-83255-1_24
Herman, G. L., & Loewenstein, J. (2017). Evidence-based change practices. *Journal of Engineering Education, 106*(1), 4–13. https://doi.org/10.1002/jee.20152
Howard, R. M. (1992). Plagiarism and the postmodern professor. *Journal of Teaching Writing, 11*(2), 233–245.
International Center for Academic Integrity (ICAI). (2021). *The fundamental values of academic integrity*. https://academicintegrity.org/images/pdfs/20019_ICAI-Fundamental-Values_R12.pdf
Jackson, E. (2025). English language support served with integrity: Reflections from a hospitality and tourism management workshop facilitator. In L. Vogt & B. M. Stoesz (Eds.), *Academic integrity in vocational and polytechnic education*. Springer.
Kessler, E. H. (Ed.). (2013). *Encyclopedia of management theory*. Sage Publications.
Koren, N. (2025). Shifting perspectives: Academic integrity in vocational education and training (VET) in the age of generative artificial intelligence. In L. Vogt & B. M. Stoesz (Eds.), *Academic integrity in vocational and polytechnic education*. Springer.
Lindstrom, G. (2022). Accountability, relationality and indigenous epistemology: Advancing an indigenous perspective on academic integrity. In S. E. Eaton & J. Christensen Hughes (Eds.), *Academic integrity in Canada*. Springer. https://doi.org/10.1007/978-3-030-83255-1_6
Lopez, S. J., & Louis, M. C. (2009). The principles of strengths-based education. *Journal of College and Character, 10*(4), 1–8. https://doi.org/10.2202/1940-1639.1041
Lorrance, A. (1974). The love project. In R. D. Kellough (Ed.), *Developing priorities and a style: Selected readings in education for teachers and parents* (2nd ed.). California State University.
Lucien-Bay, L., Couteret, L., & Radke, S. (2025). Polytechnic pinch points for academic integrity: Reflections from student conduct officers. In L. Vogt & B. M. Stoesz (Eds.), *Academic integrity in vocational and polytechnic education*. Springer.
MacIsaac, E. (2025). Exploring collusion as a form of academic misconduct in vocational shop classrooms. In L. Vogt & B. M. Stoesz (Eds.), *Academic integrity in vocational and polytechnic education*. Springer.
Miglianico, M., Goyette, N., Dubreuil, P., & Huot, A. (2021). Appreciative inquiry in the classroom: A new model for teacher development. In B. Kutsyuruba, S. Cherkowski, & K. D. Walker (Eds.), *Leadership for flourishing in educational contexts*. Canadian Scholars.
Morris, E. J., & Carroll, J. (2011). *Policy works: Recommendations for reviewing policy to manage unacceptable academic practice in higher education*. Higher Education Academy.
Morris, E. J., & Carroll, J. (2016). Developing a sustainable holistic institutional approach: Dealing with realities "on the ground" when implementing an academic integrity policy. In T. Bretag (Ed.), *Handbook of academic integrity* (pp. 449–462). Springer Singapore. https://doi.org/10.1007/978-981-287-098-8_23

Munn, C., Rexe, D., & Seeland, J. (2025). Successful institutional change: Lessons on building a culture of academic integrity. In L. Vogt & B. M. Stoesz (Eds.), *Academic integrity in vocational and polytechnic education*. Springer.

Murdoch, K., & House, D. (2024). Courageous conversations: Approaching amnesty through honesty as reparations to a learning community. In S. E. Eaton (Ed.), *Second handbook of academic integrity*. Springer Nature. https://doi.org/10.1007/978-981-287-079-7_181-1

Pang, B., Garrett, R., Wrench, A., & Perrett, J. (2018). Forging strengths-based education with non-traditional students in higher education. *Curriculum Studies in Health and Physical Education, 9*(2), 174–188. https://doi.org/10.1080/25742981.2018.1444930

Raby, R. L., & Valeau, E. J. (2018). Introduction. In R. Latiner Raby & E. J. Valeau (Eds.), *Handbook of comparative studies on community colleges and global counterparts* (pp. 3–6). Springer International Publishing. https://doi.org/10.1007/978-3-319-50911-2_1

Ratcliffe, C., Macnaughton, S., Vuori, T., & Phillips, Y. (2025). Building an academic integrity plan: Our journey at a mid-sized polytechnic. In L. Vogt & B. M. Stoesz (Eds.), *Academic integrity in vocational and polytechnic education*. Springer.

Rosenkrantz, O. (2013). The challenge of finding faculty time for applied research activities in Ontario colleges. *College Quarterly, 16*(3) https://collegequarterly.ca/2013-vol16-num03-summer/rosenkrantz.html

Royer, D. W., & Latz, A. O. (2016). Community college leadership transition through the framework of appreciative inquiry. *Community College Journal of Research and Practice, 40*(8), 695–705. https://doi.org/10.1080/10668926.2015.1072594

Sappa, V., Boldrini, E., & Barabasch, A. (2019). Teachers' resilience in vocational education and training (VET). In S. McGrath, M. Mulder, J. Papier, & R. Suart (Eds.), *Handbook of vocational education and training: Developments in the changing world of work* (pp. 1–18). Springer International Publishing. https://doi.org/10.1007/978-3-319-49789-1_28-1

Schneider, B. (2025). Advocating for fair compensation and employment equity: Reducing the burden of academic integrity and misconduct work in colleges and polytechnics. In L. Vogt & B. M. Stoesz (Eds.), *Academic integrity in vocational and polytechnic education*. Springer.

Scurr, C. (2025). Faculty education on academic integrity: A micro-credential approach. In L. Vogt & B. M. Stoesz (Eds.), *Academic integrity in vocational and polytechnic education*. Springer.

Seeland, J., Cliplef, L., Hengen, M., Joyce, K., & Murray, K. (2025). From mega to micro: Collaborative strategies for enhancing academic integrity in a college nursing program. In L. Vogt & B. M. Stoesz (Eds.), *Academic integrity in vocational and polytechnic education*. Springer.

Taylor, A. (2020). *Vocational education in Canadian schools: Toward expansive experiential learning*. EdCan Network.. https://www.edcan.ca/articles/vocational-education-in-canadian-schools/.

Urbanski, C., & Vogt, L. (2025). When technology uncovers the truth: Exploratory meetings in architectural technology. In L. Vogt & B. M. Stoesz (Eds.), *Academic integrity in vocational and polytechnic education*. Springer.

Verhoef, A. H., Fourie, M., Janse van Rensburg, Z., Louw, H., & Erasmus, M. (2022). The enhancement of academic integrity through a community of practice at the North-West University, South Africa. *International Journal for Educational Integrity, 18*(1). https://doi.org/10.1007/s40979-022-00115-y

Watts-Rynard, S. (2024, February 6). Canada's cap on international students shows a university bias. *The Globe and Mail.* https://www.theglobeandmail.com/business/commentary/article-international-students-cap-universities-graduate-programs/

Weiss, J. (2021). *The devaluation of career technical education (CTE) by the graduates of Wonderwood High School*. [Doctoral dissertation, Georgia Southern University]. https://digitalcommons.georgiasouthern.edu/etd/2290

Whitney, D. D., & Trosten-Bloom, A. (2010). *The power of appreciative inquiry: A practical guide to positive change* (2nd ed.). Berrett-Koehler Publishers.

Zhang, Y.(H.)., Lin, H., Zhang, X., & Ye, Q. (2021). The next steps in academic integrity—Education, awareness, norms, duty and law. *Forensic Sciences Research, 6*(4), 341–346. https://doi.org/10.1080/20961790.2021.1970887

Lisa Vogt is an instructor teaching ethics, leadership, and communication-focused courses in the Civil Engineering Technology department at Red River College Polytechnic in Winnipeg, Manitoba, Canada. She has dedicated over 20 years to higher education, beginning her career as an English as an Additional Language (EAL) teacher in Tokyo, Japan, before transitioning to Canadian higher education. Lisa has published and presented on academic integrity at local, national and international venues, sharing research findings and practitioner insights. The common thread in her work is creating positive and inclusive educational environments that enable students and staff to thrive. She holds a B.A. in Intercultural Studies, a certificate in Teaching English to Speakers of Other Languages (TESOL), and an M.A. in Higher Education Administration and Leadership.

Brenda M. Stoesz is the Research Lead – Science of Teaching and Learning at the Centre for the Advancement of Teaching and Learning at the University of Manitoba in Winnipeg, Manitoba, Canada. Brenda began her teaching journey in 2000, teaching high school general science and biology before moving on to teach human anatomy and physiology at Assiniboine College's Licensed Practical Nursing program. Years later, during her graduate studies in psychology, her interest in teaching led to numerous teaching assistant, laboratory, and sessional instructor positions. As a researcher, Brenda's interests vary widely from academic integrity policy analysis to effectiveness of educational interventions, particularly how technologies (e.g., virtual reality, generative artificial intelligence) can be used to promote academic integrity, prevent academic misconduct, and enhance teaching and learning.

Part I
Building and Maintaining Institutional Cultures of Academic Integrity

Chapter 2
Successful Institutional Change: Lessons on Building a Culture of Academic Integrity

Caitlin Munn, Deanna Rexe, and Josh Seeland

Abstract Cultural change is often required to support a positive academic integrity environment, a challenge for small institutions and community colleges alike. This case study demonstrates how planned, incremental changes to institutional practices and assessment approaches shifted a college's culture from rule compliance to an academic integrity culture. Employing Gallant and Drinan's (2008) four-stage model of institutionalization, this research develops lessons for other institutions, contributes to a greater understanding of academic integrity in applied education, and illustrates how the community college setting is conducive to institutionalizing academic integrity through practical processes and tools. The chapter ends with recent institutional work to further embed academic integrity into institutional quality assurance framework standards, processes, and evaluations and how the cultural evolution of quality assurance facilitates making academic integrity an integral part of all aspects of academic programming, from design to evaluation.

Keywords Academic integrity · Organizational change · Academic culture · Quality assurance · Institutionalize

Introduction

Building an institutional culture of academic integrity within post-secondary institutions is widely recognized as essential for maintaining the academic integrity and quality of education. Academic integrity culture is referred to throughout the academic integrity literature, but as Gray (2022) notes, "definitions of what such an academic culture would look like vary widely" (p. 190). For Assiniboine College (previously Assiniboine Community College; name changed to Assiniboine College

C. Munn (✉) · D. Rexe · J. Seeland
Assiniboine College, Brandon, MB, Canada
e-mail: munnc@assiniboine.net; rexed@assiniboine.net; seelandjl@assiniboine.net

L. Vogt, B. M. Stoesz (eds.), *Academic Integrity in Vocational and Polytechnic Education*, Ethics and Integrity in Educational Contexts 8,
https://doi.org/10.1007/978-3-031-90091-4_2

in June 2024), academic integrity culture is centred on the International Center for Academic Integrity (ICAI, 2021) definition of academic integrity as a "commitment to six fundamental values: honesty, trust, fairness, respect, responsibility, and courage" (p. 4). Upon these values, "a reputation of excellence and student achievement is formed" (Assiniboine Community College, 2023, p. 1). Bertram Gallant and Drinan (2008) state in reflecting on extant literature, academic integrity "can be considered institutionalized when it is integrated into organizational routines, processes, and structures, but also when it emerges as a stable norm that guides teaching and research conduct and acts as a value that binds the community" (p. 33).

As suggested by this volume, there is a need to further develop the field of academic integrity in the context of applied education and vocational education. The existing literature on academic integrity cultures in community colleges has been noted to be sparse (Hamilton & Wolsky, 2022), with most research focusing instead on universities or four-year institutions. This overlooks a unique and important context of community colleges, including differences in mandate and distinct institutional structure and dynamics and the vital role colleges play in the economic and social well-being of the communities they serve. Specifically, colleges and polytechnics have a specific purpose within higher education systems to produce graduates ready for employment, including skills, knowledge, and attitudes related to employability (Dennison, 1995; Hogan & Trotter, 2013; Jones, 2014). This mandate necessitates upholding academic integrity to ensure students are prepared to be successful in the workplace; ethical reasoning and conduct have been identified as key employability skills (Elrayah, 2021; McCowan, 2015; Reid, 2016).

This case study will explore Assiniboine College's (hereinafter Assiniboine) ten-year journey of organizational development and change to institutionalize an academic integrity culture. This journey began with a culture focused on rule compliance and, through a grassroots approach that leveraged decision-making tools and organizational structure, worked towards institutionalizing academic integrity by building a culture of academic integrity. Throughout this journey, Assiniboine anchored its approach to academic integrity in teaching and learning and focused on the factors that Bretag and Mahmud (2016) identified as contributing to exemplary academic integrity policy and the development of a culture of academic integrity.

Using case analysis, this chapter will illustrate how the community college setting is conducive to institutionalizing academic integrity culture and how particular tools, approaches, and elements of organizational structure can effectively harness that potential. Also explored is how Assiniboine's educational philosophy of *Learn by Doing* is leveraged to cultivate a culture of academic integrity and support innovation in assessment that upholds integrity and mitigates the risk of academic misconduct. Lastly, the authors will highlight Assiniboine's recent work to leverage quality assurance to make academic integrity an integral part of all aspects of academic programming, from design to evaluation, further supporting progression to institutionalizing academic integrity.

Background

Academic Integrity Culture

Institutions are increasingly required to respond to changes and pressures related to academic integrity, such as contract cheating, the sudden shift to emergency remote learning, and the rise of artificial intelligence (Kutbi et al., 2024). Historical practices and models, some built upon American research, have not fully addressed the Canadian context (Eaton & Christensen Hughes, 2022). Canadian research and efforts in academic integrity are growing in response to these (Eaton & Christensen Hughes, 2022). Researchers have found that an educational approach anchored in teaching and learning contributes to a positive, proactive, educative, and supportive culture of academic integrity (Bertram Gallant, 2008; Whitley & Keith-Speigel, 2002). However, historically, academic misconduct has been characterized as a student problem to be addressed within policy (Bertram Gallant, 2008; Whitley & Keith-Speigel, 2002). This particular organizational culture focuses on academic misconduct and rule compliance and is characterized as negative, punitive, and reactive; administrators are viewed "as the academic honesty police" (Robinson & Glanzer, 2017, p. 212). These learning environments are also where students may categorize their instructors as being rule compliance "enforcers" or "enablers" of teaching practices, which could increase academic misconduct (Robinson & Glanzer, 2017, p. 216). This is because such an environment is focused on negative messaging rather than positive messaging that motivates students and equips them with the knowledge and skills to demonstrate academic integrity and avoid academic misconduct.

Factors Supporting Academic Integrity Policy Implementation

Academic integrity policy is integral to culture building and institutional change, but it can be challenging to implement. Bretag and Mahmud (2016) specifically identify the following six factors for implementing exemplary academic integrity policy: (1) Academic integrity champions: Individuals or groups in the institution who drive change; (2) Academic integrity education for staff and students: Education for all stakeholders in the college, including students and faculty; (3) Student engagement: Students are partners in academic integrity education; (4) Robust decision-making systems: Fair and consistent responses to academic misconduct, including policy, decision-making tools, and a designated academic integrity role; (5) Record keeping for evaluation: Keeping records centrally and using the data to address issues; (6) Regular review of policy and process: Policy is an essential starting point, but it must be continually reviewed and improved based on feedback and data. At the centre of this model is a commitment to a culture of academic integrity, to which these six factors contribute.

Processes for Successful Academic Integrity Cultural Change

A culture of academic integrity manifests itself at systematic and organizational levels when viewed as an issue at this broad level (Miron et al., 2021). It is also holistic and collaborative, involving multiple stakeholder groups (Robinson & Glanzer, 2017). Given that institutional and cultural change demands a holistic transformation, involving students and instructors as well as administrative structures, policies, and institutional norms is required.

Bertram Gallant's work (2008, 2017) offers a change strategy emphasizing improvements to teaching and learning. Education is a pivotal tool in promoting integrity, where ethical conduct is learned and valued. Specific learning environments and the broader institutional environment in which they exist are the focus for continuous improvement (Bertram Gallant, 2008). The leveraging of teachable moments (Bertram Gallant, 2017) supports a move away from rule compliance towards the building of ethical knowledge through institutional support, supporting the development of a culture of academic integrity. Grassroots approaches are more successful than administrative directives (Bertram Gallant, 2011), and policy, too, is important in an academic culture (Bretag & Mahmud, 2016). Students also respond more positively to grassroots initiatives; they disapprove of academic integrity being "something enforced from the top down through administrators and teachers (Robinson & Glanzer, 2017, p. 213).

Case Study at Assiniboine

The case is Assiniboine's journey of institutionalizing a culture of academic integrity over 10 years, from 2013–2023 (see Appendix A for a detailed timeline). A case study methodology was employed (Yin, 2014) because it supports an in-depth exploratory approach to a new area of enquiry, specifically to understand the activities, events, and behaviours that facilitated specific institutional change. The case was developed following an extensive search in which data were gathered from college documents, including academic integrity policies and procedures, briefing notes, scoping reviews, Academic Integrity Advisory Committee (AIAC) reports and work products, training and professional development materials, student materials, Library Guides, institutional standards and processes, job descriptions, and Assiniboine participation in professional community activities. These data were used to construct a historical timeline involving key events, actors, and contexts; 52 unique events were identified and described. Trustworthiness of this case is developed through the involvement and agreement of key actors; the three co-authors are employees with direct knowledge of this history, having been directly involved during this ten-year period.

Bertram Gallant and Drinan's (2008) four-stage model of institutionalization has been selected as the overall conceptual framework for analyzing this case because of its specific focus on academic integrity; it allows us to examine how an institution can "mobilize around a change initiative or innovation, implement that innovation, and then see the innovation become stabilized or institutionalized within the organization" (Bertram Gallant & Drinan, 2008, p. 29). The four stages of institutionalization are recognition and commitment, response generation, response implementation, and institutionalization. The first step is for the institution to identify the key issue and "commit to addressing that issue with organizational change" (p. 31). Once the key issue is identified, a response is generated and implemented, necessitating changes within the institution to support new practices. Academic integrity is institutionalized when it becomes embedded into the institution and part of the organizational practice and structure. Academic integrity becomes a shared part of the institution's culture, guiding teaching and learning.

Bertram Gallant and Drinan (2008) use a pendulum as a metaphor for this model to show the "fluidity and movement within and between stages" (p. 29). This metaphor illustrates that the stages are not linear; however, they provide an organizational feature to map out historical events and undertake this case analysis. Assiniboine has actively used scholarly literature on academic integrity and membership in scholarly communities to develop institutional responses. Bertram Gallant and Drinan's (2008) four-stage model of institutionalization has served as a guide to benchmark institutional change in academic integrity. While the model is valuable, it is important to recognize that not all its aspects apply to Assiniboine and potentially the broader Canadian context. Assiniboine did not adopt honour codes as part of response generation, which Bertram Gallant and Drinan (2008) identify as a consideration for institutions. Honour codes, which feature more prominently in American higher education, have not been widely adopted in the Canadian context (Eaton & Christensen Hughes, 2022).

In developing responses to the changing environment, since 2013, Assiniboine has intentionally incorporated what is applicable to our context rather than applying any model entirely, as is the case with Bertram Gallant and Drinan's (2008) model. In doing so, Assiniboine has placed itself and its academic integrity culture in the broader context of higher education while developing its own identity. This identity is in accordance with the idea that "institutions can establish what works in other organisations and apply or adapt that practice to their own situation" (QAA, 2022, p. 42).

Case Description

Assiniboine is a medium-sized publicly funded community college in Manitoba, Canada, and it has permanent campuses across the province, including Brandon, Dauphin, Southport, and Winnipeg. Uniquely, Assiniboine also partners with dozens of communities and organizations in Manitoba to deliver community-based programs. By legislation, Assiniboine "has a responsibility to enhance the economic and social well-being of Manitoba through the provision of a broad range of educational opportunities" (The Colleges Act, C.C.S.M. c. C150.1, 2024). In accordance with this mandate, Assiniboine offers applied programs in a variety of disciplines, including culinary arts and hospitality, business, agriculture and environment, health and human services, and trades and technology. Assiniboine also delivers apprenticeship programming in a range of skilled trade disciplines. Assiniboine aims to graduate students with knowledge, skills, and abilities that prepare them for career-related employment and employs an educational philosophy of *Learn by Doing,* in which students integrate knowledge through experiential learning.

Before actively developing an academic integrity strategy in 2013, Assiniboine's approach to academic integrity was negative, punitive, and reactive. For example, the Library and Learning Curve (student academic support centre) offered in-class sessions for students on writing, citing, and plagiarism. These sessions were reactive because instructors scheduled them in response to academic misconduct incidents in their courses. Whereas proactive sessions are provided early in the course or program, with the goal of promoting academic integrity and reducing academic misconduct before it occurs.

Responsive to changes in the learning environment, in 2015, the authors worked together to begin offering sessions on academic integrity more broadly, timed to precede or coincide with typical assignment deadlines, with more focus on building academic skills and the importance of integrity rather than the problems with academic misconduct. Following the introduction of international student cohorts, multiple Assiniboine stakeholders identified a need for a proactive response. The Library initiated an informational briefing regarding international students and the potential for plagiarism concerns to support proactive planning and identify factors influencing plagiarism among international students. As a result, in 2016, the Dean of Students committed to a working group on plagiarism. Two of the authors chaired subcommittees for this working group.

In 2016, the plagiarism working group's focus was broadened to include members of three subcommittees, each representing a different student demographic—international, domestic, and Indigenous students. The working group included 16 staff and faculty from 7 programs, 3 schools, and 3 campuses. Later that year, the subcommittees met to share findings and identify commonalities. Results were presented to the Dean of Students, and an action plan was created based on the findings. This included designing and delivering collaborative sessions that brought together research and writing to provide proactive skill development on citing and avoiding plagiarism. A presentation was provided to the Learning Leadership Council (the academic coordinating body) on the findings of the working group with the goal of supporting action in schools and programs. The report was favourably received.

As a result, in 2017, Assiniboine increased its focus on studying and understanding the academic misconduct issues faced by campus stakeholders. To support this work, Assiniboine broadened from the internal college perspective alone to the provincial, national, and international landscapes. At this time, Assiniboine joined ICAI to gain access to resources, including the self-assessment tool, and further explore the values of academic integrity as being a core part of Assiniboine's response. Assiniboine staff also attended the first Academic Integrity Inter-Institutional Meeting (AIIIM), founded by the Centre for the Advancement of Teaching and Learning at the University of Manitoba, an event for post-secondary schools in Manitoba to share and collaborate on academic integrity in teaching and learning. Assiniboine's commitment led to a new position, a student peer tutor for academic integrity. The peer tutor brought a student perspective and peer support approach to in-class workshops on academic integrity and one-on-one support for students. The peer tutor also assisted in creating resources to help students avoid academic misconduct. At this time, one of the authors also created an academic integrity map with an intercultural focus on the six fundamental values of academic integrity (ICAI, 2021). The purpose was to build student congruence around academic integrity rather than a false dichotomy about academic misconduct. The map demonstrated the universality of these values, providing examples from different historical and cultural contexts to show that academic integrity transcends time and culture and is a shared human value. This approach encouraged viewing academic integrity beyond the limited perspective of specific cultures or time periods.

In 2018, the Learning Commons merged with the Centre for Learning Innovation (CLI, the teaching and learning centre), creating a comprehensive integrated unit supporting both faculty and students. Academic integrity initiatives were primarily spearheaded by the Learning Commons, with interprofessional collaborative teamwork within the Learning Commons providing services which enhance teaching and learning for all the members of Assiniboine. As an early priority, the group targeted assessments that historically saw challenges, including higher rates of academic misconduct, and redesigned them with the instructor to reduce opportunities for academic misconduct. The authors also began offering interactive sessions for academic leaders and faculty on APA citations to calibrate APA expectations and address citation fixation. Assiniboine also hosted the second annual AIIIM this year.

The Learning Commons became increasingly involved in academic integrity work, directly providing faculty support, assessment redesign support, professional development on academic integrity, and more. To inform this approach, the Learning Commons utilized the ICAI self-assessment tool to calibrate institutional progress and develop departmental planning priorities. The authors also presented at provincial conferences with a focus on sharing faculty outreach opportunities and ways Assiniboine integrated academic integrity into student transition and orientation.

In 2019, Assiniboine also began to develop a comprehensive vision that brought together various academic integrity frameworks (Bertram Gallant, 2017; Stephens, 2016; Quality Assurance Agency for Higher Education, 2022) to conceptualize an institutional approach to academic integrity of education, prevention, identification, and administration (see Fig. 2.1). The model was used internally to guide our work and educate stakeholders. Meaningfully, the dominant strategy was educative and not punitive, a strategy which was generally welcomed on campus.

Assiniboine also launched the Assiniboine AIAC to engage a wider group of campus stakeholders more formally, following an existing model of campus engagement used by the Library. The committee had members from six different Assiniboine campuses and sites, meeting quarterly. AIAC served three main purposes: (1)

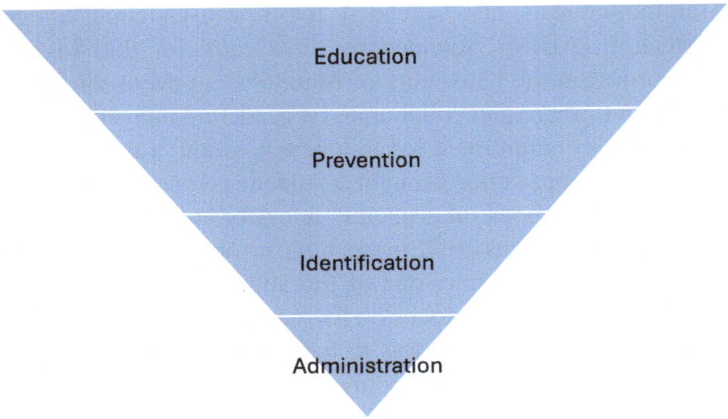

Fig. 2.1 Assiniboine's College's institutional approach to academic integrity and academic misconduct. *Note*. From "Academic integrity for faculty library guide," by Assiniboine College, n.d., https://assiniboine.libguides.com/ai-faculty#s-lg-box-16368504

informing services, projects, and resources related to academic integrity in the Learning Commons; (2) identifying opportunities for implementing and evaluating academic integrity interventions and supports; and (3) contributing to the development of policies and procedures.

In 2019, Assiniboine also expanded professional development for faculty on assignment design and integrated academic integrity topics into the mandatory Teaching Essentials Program for new instructors. Assiniboine also held events for the International Day of Action Against Contract Cheating and joined the Manitoba Academic Integrity Network (MAIN; Stoesz et al., 2020), furthering its participation with coalition partners to advance academic integrity in the public post-secondary sector. Additionally, an online academic integrity module was developed within the college's learning management system (LMS, i.e., Moodle) to be proactively integrated into student orientation. The module introduced the values of academic integrity, outlined expectations according to the college academic integrity policy, highlighted available college resources and support for students, and assessed student understanding through a series of case study questions. Lastly, a briefing note on contract cheating, rephrasing, and file-sharing websites was initiated by the Library as a response to increasing concerns and was presented to academic decision-makers; this led to a decision to block websites facilitating academic misconduct from being accessible on the Assiniboine network, as part of a multi-institutional collaboration (Seeland et al., 2020).

In 2020, Assiniboine initiated a formal renewal of the academic integrity policy, which had been used since 2013. However, due to the disruptions caused by the COVID-19 pandemic, the policy renewal was slower than anticipated, although the shift to remote and online instruction created motivation and a need to ensure academic integrity in these settings. Assiniboine quickly leveraged academic integrity champions from the AIAC and the Learning Commons to provide academic integrity education for staff and students and focus on related policies and procedures. Urgently required decisions on emergent issues were supported by scoping reviews and briefing notes, which had become standard practices at the college. One example was the college's policy decision to forbid the use of e-proctoring; a briefing note informed this decision.

In 2021, academic integrity began to be integrated into Assiniboine's quality framework by creating Learning Management System (LMS) Minimum Presence Standards. By outlining standards for LMS course pages and teaching, Minimum Presence Standards provide a consistent student experience across their courses and ensure that students can access the critical information and curriculum assets they need to succeed. The standards also guide faculty on good practices in course design, development, and delivery. Integral to the development of these standards were considerations of the factors that influence academic integrity and academic misconduct, with a strong focus on the role satisfaction or dissatisfaction with the learning environment plays (Bretag et al., 2022).

Given the new instructional practices that had emerged during the COVID-19 pandemic and increased faculty fluency in educational technology, there was a notable increase in requests from faculty for plagiarism detection tools. The Library responded to these pressures again by reviewing literature and best practices, resulting in a briefing note on text-matching software. The briefing note focused on the importance of using tools for education and prevention rather than reactive detection. Software options were identified and evaluated, which led to a recommendation for software which searches for direct matches between student work and content available online and within the institutional repository of previous student submissions. This tool provides a "similarity score," a summary of sources the submission matched, a list of matches, and a comparison between the submitted student work and the content that matches. The investigative work supported the institutional decision to obtain a subscription to Urkund/Ouriginal. To support a positive implementation, a faculty training module was developed to use the tool in a manner that aligned with the college's positive, proactive, and supportive approach to academic integrity.

In 2021, mandatory academic integrity education for all first-year students became embedded into a required course. This course, developed for all regular college programs, integrates aspects of college orientation and policy education and aims to equip students with the skills and strategies to navigate college expectations successfully. Academic integrity is articulated in a course learning outcome and has an aligned assessment. This module was further expanded in 2022 to incorporate specific education on generative artificial intelligence.

Based on feedback from the committee, the Learning Commons was prompted to develop more comprehensive faculty support in 2022. Supportive job aids and professional development sessions were developed based on assessment design strategies gathered from the literature, including the development of higher order thinking skills, individualizing assignments, avoiding commercial test banks, and regularly refreshing assessments. Further, the authors and members of the Learning Commons broadened their work to examine academic integrity in specific disciplines such as math (Seeland et al., 2022), enabling them to support the various programs across the college better.

Following input from an extensive college consultation in 2023, the new Policy A25: Academic Integrity completed its renewal process. The prior policy was published in 2013, used outdated vocabulary to describe academic misconduct, and did not represent Assiniboine's current responses. The policy language was updated to reflect current forms of academic misconduct (e.g., from essay mills to contract cheating sites) and to remove words that reinforced a moralistic and punitive approach, such as dishonesty, evidence, and guilt. Also implemented was a shift from a *burden of proof* assumption to a *balance of probabilities*, aligning with campus norms in administrative justice in decision-making. The policy supported campus members in using information-gathering strategies such as metadata and bibliographic analysis, academic judgment based on knowledge of students, their previous work, and a discovery interview. The Library developed a procedures manual that outlined steps and recommended sanctions, along with definitions and tools to support multiple stakeholders and decision-makers.

Analysis

Using Gallant and Drinan's (2008) four-stage model of institutionalization, Fig. 2.2 summarizes and maps key institutional steps.

Pre-Recognition and Commitment (2013–2015)

During this phase, initial awareness and preliminary initiatives were developed to address academic misconduct, focusing primarily on rule compliance. Key events included the introduction of workshops and student support for academic integrity, APA citation, and writing skills. In 2015, a proactive report on plagiarism prevention was produced, which served as a critical trigger for advancing to the next stage.

Recognition and Commitment (2016)

Assiniboine formally acknowledged the need for a structured approach to academic integrity. This stage saw the formation of a working group on plagiarism, proactive workshops on citing and plagiarism, and training for academic managers, including briefing note training and principles of administrative justice training. These efforts marked the beginning of a shift from a rule compliance culture to one focused on academic integrity, setting the foundation for generating effective responses to academic integrity challenges.

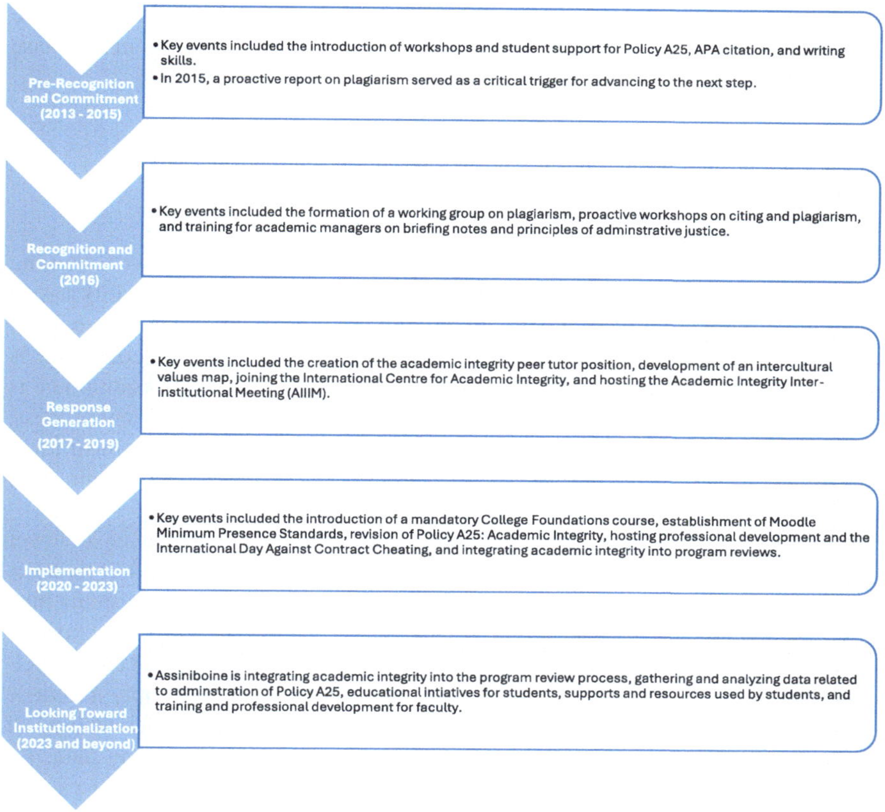

Fig. 2.2 Key institutional steps to promoting academic integrity

Response Generation (2017–2019)

Between 2017 and 2019, Assiniboine generated various responses to support academic integrity, including the creation of the peer tutor role, the development of an intercultural values map, joining the ICAI and hosting multiple AIIIMs. New policies on standardizing citation style and various faculty support initiatives were introduced, demonstrating a growing commitment to embedding integrity into institutional practices. These efforts provided a foundation for the next phase, moving from generating responses to actively implementing them across the institution.

Implementation (2020–2023)

From 2020 to 2023, Assiniboine focused on implementing responses, including introducing a mandatory course, establishing Moodle Minimum Presence Standards, revising the academic integrity policy, faculty training, hosting professional development events, and integrating academic integrity into program reviews. These actions demonstrate the institution's commitment to transitioning from response generation to full implementation.

Looking Toward Institutionalization (2023 and Beyond)

As of 2023, Assiniboine is partway through the Implementation stage, with ongoing efforts to fully institutionalize academic integrity by leveraging the quality assurance framework to embed academic integrity into all aspects of academic programming, from design to evaluation. While the full institutional data collection is incomplete, there is data collection at the school and program level. Assiniboine is integrating an examination of academic integrity into the program review process, gathering and analyzing data related to academic integrity policy administration, educational initiatives for students, student supports and resources, and training and professional development for faculty.

Institutions at the stage of institutionalization may also aim to have fewer instances of academic misconduct (Gallant & Drinan, 2008); however, our data has not shown a decrease. As Kenny and Eaton (2022) pointed out, new systems may cause a temporary increase in reported incidents. Therefore, we did not expect a quick reduction in academic misconduct cases while the culture was being developed and implemented; culture change takes time. Our goal is not simply to reduce the number of academic misconduct cases but to foster a deeper understanding of the factors contributing to academic misconduct at the institution using data. By capturing and analyzing data, we can work to identify patterns and root causes, enabling us to implement targeted interventions and continuous improvements. This is part of a larger institutional approach aiming to cultivate an academic integrity culture.

Next Stage of Development

Assiniboine is working to further embed academic integrity into its institutional quality assurance framework policy document, including within standards, processes, and procedures. As Thacker and McKenzie (2022) explained:

> quality assurance processes may not only be an effective tool for managing and assuring program quality but can also be leveraged to support the continuous improvement of policies and practices and the development of cultures, in keeping with academic integrity. (p. 529)

Assiniboine seeks to leverage this intersection between quality assurance and academic integrity to make academic integrity an integral part of all aspects of academic programming—design, development, delivery, and evaluation. This is timely work, given that Assiniboine is in the process of developing its institutional quality framework and building a quality system that operationalizes new processes and tools while renewing existing ones. A recent action explicitly addressed our approach to academic integrity as a standard within the Academic Charter as part of its renewal. The Academic Charter outlines Assiniboine's applied education philosophy, program standards, and key practices. Assiniboine's Academic Charter (2024) now includes the standard: "A proactive, educative, and supportive approach to academic integrity is utilized to support students to learn with integrity while protecting the integrity of assessment and evaluation."

Assiniboine is also integrating an examination of academic integrity into the program review process, inspired by Thacker and McKenzie (2022). This involves gathering and analyzing data related to Policy A25: Academic Integrity administration, as well as evaluating the program's educational initiatives for students, supports and resources used by students, and training and professional development for faculty. The collected data is then used to identify areas in the curriculum that could be improved to mitigate the risk of academic misconduct, such as assessments with higher incidents of misconduct. Furthermore, this data helps tailor student education initiatives and enhance faculty professional development to address identified gaps. Insights gathered through program review will inform the renewal of program curriculum, ensuring that academic integrity is continuously monitored and strengthened across all programs as they undergo review.

This quality assurance work builds upon the events and activities in 2020–2023 that integrated academic integrity into program, course, and assessment design, institutionalizing it into quality assurance standards, processes, and procedures. For example, academic integrity was intentionally integrated into the course design process at Assiniboine and within the training offered to faculty. The Course Design: Blueprinting course, which teaches faculty how to design a course, includes intentionally addressing academic integrity in the course design process. This is crucial because it ensures academic integrity is embedded in the curriculum from the outset. This proactive approach helps prevent academic misconduct by creating clear expectations and guidelines for students and developing assessments that are less susceptible to academic misconduct. Furthermore, incorporating academic integrity into faculty training for course design ensures that instructors are well-equipped to model integrity for their students and create a learning environment with a positive, proactive, and supportive approach to academic integrity.

Assiniboine experience highlights the importance of including academic integrity in quality assurance. By integrating academic integrity into program reviews and other quality assurance activities, institutions can identify areas for improvement and enhance both student education and faculty development in academic integrity.

Discussion

Five key features of Assiniboine's success are identified through this analysis, some consistent with the literature and some new additions.

Institutional Capacity and Process for Advancing Issues for Decision

Progress was aided by the employees' ability to articulate an issue to the appropriate decision-maker and to influence and facilitate the decision-making process. Formal staff training in developing briefing notes as the institutional standard and expectation for decision-making were crucial tools for advancing academic issues for decision; the college instituted this practice in 2012. In the examples described in this case, standardized problem analysis, option identification, and evaluated options were key in articulating issues, providing a background, and outlining options ready for decision.

Organizational and Governance Structure & Access to Decision Makers

Community colleges may have advantageous environments conducive to creating cultures of academic integrity. In our experience, identifying a decision-maker is facilitated by Assiniboine's organizational structure and governance processes. The relatively flat and streamlined organizational structure allows staff to efficiently bring issues to the appropriate decision-makers, which also, due to the policy environment, are clear to campus members. Clear knowledge of decision-makers aids in preparing effective decision notes, which, as an established college process, supports informed and responsive decision-making by academic leaders. These factors, combined with supportive organizational leadership, have enabled the authors and the Assiniboine's AIAC to effectively address academic issues, thereby advancing innovation, change, and the culture of academic integrity.

Additionally, the organizational and governance structure of the college allowed for a distributed leadership model through committees and workforces on particular topics or issues. The AIAC was impactful institutionally because it allowed for knowledge of academic integrity "within and across units" (Kenny & Eaton, 2022, p. 587). This assignment of responsibility and authority provided for authentic and meaningful stakeholder engagement and genuine collaboration at the grassroots level. The committee's membership and terms of reference supported a coordinated group working together in pursuit of a culture of academic integrity. Individuals across units in the college came together to bring forth issues and possible solutions.

Integrated Faculty and Student Learning Commons Supports

Providing teaching and learning supports, including academic integrity for staff and students, is key to fostering academic integrity cultures (Bretag & Mahmud, 2016). Assiniboine's organizational structure facilitates this by reducing silos and hierarchy, allowing for quick action to emerging issues and clear communication between stakeholders involved. For example, the proximity and collaboration between units in the Learning Commons enable swift responses to academic issues, including offering assessment redesign support and professional development on academic integrity. This interdisciplinary approach has established the unit and its members as continuous contributors to the culture of academic integrity.

Several job descriptions within the Learning Commons now explicitly outline responsibilities related to academic integrity support. Awareness and support for academic integrity and a clear process for updating job descriptions were integral to accomplishing this. Senior leadership recognized the importance of the unit for institutional academic integrity efforts and intentionally incorporated these responsibilities when creating new positions or renewing job descriptions. This proactive approach, supported by a streamlined process for managers, ensured that several roles within the Learning Commons now explicitly focus on academic integrity support. As a result, the Learning Commons and its members are integral to providing aligned and intentional support for both faculty and students, facilitating swift responses to academic issues, and promoting a cohesive, academic integrity-focused approach. The Learning Commons and AIAC were integral in informing decision-making on academic integrity and the provision of support during the disruptions caused by the COVID-19 pandemic.

Anchoring Decisions in Teaching and Learning

Like many other institutions, in March 2020, Assiniboine was forced to adapt practices and assessments for emergency remote instruction quickly. Even during emergency responses, the college was disciplined in decision-making to ensure alignment with the Academic Charter and college philosophy of education. As a result, pandemic-related decisions supported high-quality teaching and learning environments, taking into consideration the factors that influence academic integrity and academic misconduct, such as satisfaction with the learning environment and the perception that there are opportunities to cheat (Bretag et al., 2022). Assiniboine leveraged its approach and culture of assessment to continue to anchor decisions in teaching and learning rather than focusing on rule compliance.

Philosophy of Education and Focus on Vocational Education

Because Assiniboine aims to provide education that is aligned with industry needs and produce graduates ready for employment, learners must gain the skills and knowledge required by business and industry. This influences Assiniboine's evaluation culture to prioritize the authentic and practical demonstration of knowledge and skills over more traditional standardized tests and exams; "assessments and evaluations are authentic in that they measure performances that are as close as possible to real-world experiences and applications" (Assiniboine College, 2024). This philosophy of education allows Assiniboine to leverage assessment practices that support cultures of academic integrity and reduce avenues for academic misconduct. Thus, as an applied education institution, our approach of *Learn by Doing* further supported the institutionalization of academic integrity at Assiniboine by prioritizing a proactive and educative academic integrity culture over rule compliance and academic misconduct.

Conclusion

Analyzing how Assiniboine transitioned from rule compliance towards institutionalizing a culture of academic integrity provides insights into the institutional and cultural factors at play. Applying a theoretical framework to the case allows us to explain the underlying principles, factors, behaviours, and mechanisms that drove Assiniboine's institutional and cultural change.

Consistent with the literature, academic integrity champions (Bretag & Mahmud, 2016), education (Bertram Gallant, 2008, 2017; Bretag & Mahmud, 2016), decision-making (Bretag & Mahmud, 2016), and engagement of multiple stakeholders (Robinson & Glanzer, 2017) were important factors to Assiniboine building a culture of academic integrity. Also consistent with the literature, institutional change was possible through taking an educative approach anchored in teaching and learning (Bertram Gallant, 2008; Whitley & Keith Speigel, 2002), leveraging grassroots approaches (Bertram Gallant, 2011), and policy (Bretag & Mahmud, 2016). Novel to Assiniboine was interprofessional roles working within a unit to advance academic integrity, the role of an institutional academic integrity committee, decision-making tools and access to decision-makers, and an educational philosophy that supports authentic assessment practices.

Assiniboine has moved from a punitive rule compliance culture towards institutionalizing an academic integrity culture (Gallant & Drinan, 2008). This was possible by leveraging the elements of community colleges that are conducive to a culture of academic integrity and the factors that support this cultural development (Bretag & Mahmud, 2016), as well as implementing consistent, incremental changes that gained momentum and inspired further change. We know from literature and experience that there is no one silver bullet to stop all academic misconduct and push the culture of integrity to the finish line. Rather, constant, organized, daily actions in multiple roles create cumulative change. At an institutional level, this is in line with Dawson's (2021) advice that "Perfect is the enemy of good, so aim for incremental improvements" (p. 143).

Appendix a Timeline of Key Events in Institutionalizing Academic Integrity

Year	Event	Initiator	Why?
2013	Briefing note training begins for all academic managers.	President	Identified need to articulate and document decisions at the college.
	New college Strategic Direction.	President	Routine strategic planning initiative triggered by the new president resulted in the college's goal of doubling the number of Assiniboine graduates by 2025.
	Workshops and student support on Policy A25 and APA citation are provided.	Library	Part of the Library's mission is to support students in accessing and using information.
	Workshops and student support for writing.	Learning Curve (student academic support)	Part of Learning Curve mission to support students in the development of academic skills.
2015	Report on plagiarism and how to proactively address it.	Library	In response to the announcement of planned international student cohorts.
2016	Collaboration between the Learning Curve and Library to deliver proactive workshops on citing and plagiarism.	Library & Learning Curve	Desire to proactively address plagiarism, reduce reactive response, and embed writing and citing in context.
	Working group on plagiarism. Three subcommittees, with a total of 12 administrators, support staff, and faculty.	Dean of Students (Champion) Library & Learning Curve chaired and led subcommittees	In response to the plagiarism report in 2015.
	Presentation to Learning Leadership Council on findings of plagiarism working group.	Library & Learning Curve	To share findings and support implementation of findings in school and programs areas.
	Principles of Administrative Justice training for all academic managers.	Vice President Academic	Identified need for good decision-making.

(continued)

Year	Event	Initiator	Why?
2017	Indigenization Strategy.	Vice President Academic and Dean of Students	Identified need for increased clarity of direction and purpose. Outlines strategies for community engagement to support increased enrolment and improved student success.
	Academic Plan.	Vice President Academic	To further progress towards the Strategic Plan and provide value to college planning processes through articulating an academic vision and focus, with identified academic priorities, goals and strategies for what we want to accomplish in the next several years to 2025.
	Joined the International Center for Academic Integrity (ICAI).	Learning Commons	To gain access to resources, including a survey tool that could be used to benchmark our approach to academic integrity.
	New academic integrity peer tutor position.	Learning Commons	To support student awareness of academic integrity. Peer tutor joined class workshops, developed resources, and answered student questions.
	Intercultural values map that identifies commonalities between academic integrity values and values from cultures around the world.	Learning Commons	To support a diversifying student population and make academic integrity values relevant to students.
	Attended the first Academic Integrity Inter-Institutional Meeting (AIIIM).	Learning Commons	Interested in meeting representatives from institutions across the province for professional development and collaboration on academic integrity.
	New Policy on Standardizing Citation Style introduced.	Library, Vice President Academic	Academic citation is a necessary part of academic honesty and integrity but is often a cumbersome process for students, faculty, and staff who help students. At Assiniboine, multiple citation styles were used, often necessitating students to learn and use various styles and requiring additional staff support. This policy chose one standard citation style for the college, simplifying the citation process for all stakeholders and allowing students to concentrate on developing other crucial skills.

2018	Strategic Direction refreshed.	President	Introduces three pillars: 1) an unparalleled student experience, 2) quality programs responsive to Manitoba's labour market and 3) Outreach and engagement to support economic growth.
	Centre for Learning Innovation (CLI) merges with Learning Commons.	Vice President Academic	To offer seamless, integrated services for faculty and students through one unit.
	Sessions held with faculty on APA citations and avoiding citation fixation.	Learning Commons	In response to varied APA citation expectations across the college, resulting in increased demands for student support on citing.
	Institutional posters created for each of the ICAI fundamental values of academic integrity.	Learning Commons	To build awareness on campus through posters and provide faculty with resources that could be integrated into lessons and classroom activities.
	Hosted and presented at the second Academic Integrity Inter-Institutional Meeting.	Learning Commons	To continue to support cross-institutional collaboration and learning.
	Presented at the Manitoba Library Association Conference on outreach opportunities in academic integrity.	Learning Commons	To engage with the Library community on academic integrity.
	New faculty support focused on academic integrity (e.g., assessment redesign and proactive strategies).	Learning Commons	To offer integrated faculty support by multi-disciplinary staff in the Centre for Learning Innovation and the Learning Commons.
	Review of orientation practices.	Vice President Academic	To support continuous improvement in orientation practices and identify new strategies for supporting student transition, success, and engagement throughout the college experience.

(continued)

Year	Event	Initiator	Why?
2019	Professional development offered on reducing plagiarism through assessment design, including workshops.	Learning Commons	To further scale support through group-based activities, such as workshops.
	URL blocking for contract cheating, academic file-sharing, and rephrasing sites.	Learning Commons	In response to literature and recommendations emerging on contract cheating, Assiniboine adopted URL blocking.
	Presented at the third AIIIM.	Learning Commons	To continue to support cross-institutional collaboration and learning.
	Manitoba Academic Integrity Network (MAIN) established, and Assiniboine became an inaugural member.	Library	Provincial colleges and universities decided to formalize the community developed through AIIIM and collaborate to deliver professional development throughout the year (not just once a year like AIIIM).
	Student module on academic integrity developed for use in orientation.	Learning Commons	To proactively provide education on academic integrity for all students.
	Academic integrity incorporated into the Teaching Essentials Program.	Learning Commons	To proactively educate instructors on academic integrity and how to prevent academic misconduct.
	Hosted professional development for the International Day Against Contract Cheating.	Learning Commons	In alignment with events being held internationally and nationally.
	Academic Integrity Advisory Committee established.	Library, Vice President Academic	To move from silos addressing the topic to a coordinated group working together in pursuit of a culture of academic integrity.
	Presented at the Library 2.0: Shaping the Future of Libraries with Instructional Design Conference on collaborating to offer academic integrity services and supports across campuses and locations.	Learning Commons	To continue to engage in the academic integrity community.

2020	Briefing note on e-proctoring, identifying risks and issues in its implementation.	Vice President Academic	At the request of the VPA, Learning Commons completed a report, review, and recommendations, adopted in full as policy by the Deans Council.
	Hosted professional development for the *International Day of Action Against Contract Cheating*.	Learning Commons	In alignment with events being held internationally and nationally.
	Briefing note on tracking academic misconduct data. Outlines the option of using the student information system.	Library and Registrar's Office	To support better data collection and analysis, as well as informed decision-making and continuous improvements in programs.
	Publications on academic integrity, co-written with colleagues from other Manitoba institutions.	Library	To share lessons from cross-institutional collaborations and efforts to block contract cheating sites.
2021	Development of mandatory College Foundations course for all regular college programs. Includes learning outcomes and a module specific to academic integrity.	Vice President Academic	The course was developed as an institution-wide response to student learning needs during the COVID-19 pandemic. The course is delivered online to ensure academic continuity during different phases of health safety measures.
	Further publication and conference presentations on collaboration in academic integrity, policy revision, math in academic integrity, championing academic integrity initiatives, and academic integrity culture development.	Learning Commons	To continue to contribute to the literature on academic integrity.
	Briefing note on text-matching software.	Learning Commons	To respond to requests for plagiarism detection tools in an informed manner that aligns with the college's approach to academic integrity.
	Moodle Minimum Presence Standards established, which include consideration of learning environments and factors that support quality in teaching and learning and student satisfaction.	Vice President Academic	To support quality learning through the learning management system presence as part of quality framework development.
2022	Generative artificial intelligence added to academic integrity module in College Foundations course.	Learning Commons	To keep the curriculum current, relevant, and responsive.
	Continued conference presentations and publications on contract cheating, critical thinking, and addressing academic misconduct.	Learning Commons	To continue contributing to academic integrity literature and support provincial activities.

(continued)

Year	Event	Initiator	Why?
2023	Revision of academic integrity policy led by the Academic Integrity Committee and through consultation with the college community.	Library	The policy was outdated and due for review.
	Assessment security scoping review.	Learning Commons	To support informed, research-based decision-making about the implementation of lockdown browsers. After the briefing note, the Learning Commons decided not to support some uses of lockdown browsers due to the challenges and limitations of the software.
	Presentations at conferences on evaluation practices and academic integrity, as well as establishing a culture of academic integrity.	Learning Commons	To continue contributing to the academic integrity community and at the request of other institutions as part of their professional development.
	Academic Charter update incorporates academic integrity standard.	Vice President Academic	To formalize academic integrity as part of the college academic standard and operationalize a positive, proactive, and supportive approach to academic integrity.
	Integration of academic integrity into course design and development process.	Learning Commons	To continue operationalizing academic integrity as part of quality assurance activities and processes.
	Institutional statement on artificial intelligence developed.	Vice President Academic	To provide direction on decisions about integrating artificial intelligence skills, knowledge, and attitudes into curriculum and to support curriculum innovation and maintenance. Also supports the incorporation of occupationally safe, ethical, and suitable use of artificial intelligence tools as part of our occupationally relevant training.
	Academic integrity and misconduct rates looked at during program review as part of data collection.	Learning Commons	To continue operationalizing academic integrity as part of quality assurance activities and processes.
	Teaching and learning with generative artificial intelligence guide created.	Learning Commons	To expand discussion of artificial intelligence further than use and misuse (academic misconduct). To further support programs and faculty in addressing this as a teaching and learning matter, not just misconduct. As well as to provide up-to-date information on artificial intelligence and emergent guidance (e.g. course outline statements about artificial intelligence).
	Six roundtable discussions and professional development sessions on GenAI were offered.	Learning Commons	To support dialogue with faculty on the topic.

References

Assiniboine College. (2023). *Policy A25: Policy on academic integrity.* https://assiniboine.net/sites/default/files/2023-07/A25%20Policy%20on%20Academic%20Integrity.pdf

Assiniboine College. (2024). *Academic charter.*

Bertram Gallant, T. (2008). *Academic integrity in the twenty-first century: A teaching and learning imperative.* Jossey-Bass.

Bertram Gallant, T. (2011). Building a culture of academic integrity: Based on the Magna online seminar, *"Helping students learn from ethical failures"* [White Paper]. Magna Publications, WI.

Bertram Gallant, T. (2017). Academic integrity as a teaching & learning issue: From theory to practice. *Theory Into Practice, 56*(2), 88–94.

Bertram Gallant, T., & Drinan, P. (2008). Toward a model of academic integrity institutionalization: Informing practice in postsecondary education. *Canadian Journal of Higher Education, 38*(2), 24–43. https://doi.org/10.47678/cjhe.v38i2.508

Bretag, T., Curtis, G., McNeill, M., & Slade, C. (2022). *Academic integrity in Australian higher education: A national priority.* Tertiary Education Quality and Standards Agency (TEQSA). https://www.teqsa.gov.au/sites/default/files/2022-10/academic-integrity-infographic.pdf

Bretag, T., & Mahmud, S. (2016). A conceptual framework for implementing exemplary academic integrity policy in Australian higher education. In T. Bretag (Ed.), *Handbook of academic integrity* (pp. 463–480). Springer Singapore.

Dawson, P. (2021). *Defending assessment security in a digital world: Preventing e-cheating and supporting academic integrity in higher education.* Routledge.

Dennison, J. D. (Ed.). (1995). *Challenge and opportunity: Canada's community colleges at the crossroads.* UBC Press.

Eaton, S., & Christensen Hughes, J. (2022). Academic integrity in Canada: Historical perspectives and current trends. In S. E. Eaton & J. Christensen Hughes (Eds.), *Academic integrity in Canada. Ethics and integrity in educational contexts* (Vol. 1, pp. 3–24). Springer. https://doi.org/10.1007/978-3-030-83255-1_1

Elrayah, M. (2021). Can business ethics-based education improve HR graduates' employability, knowledge, skills, and competence? *Journal of Education for Business, 96*(8), 522–529.

Gray, B. (2022). Ethics, EdTech, and the rise of contract cheating. In S. E. Eaton & J. Christensen Hughes (Eds.), *Academic integrity in Canada. Ethics and integrity in educational contexts* (Vol. 1, pp. 189–201). Springer. https://doi.org/10.1007/978-3-030-83255-1_9

Hamilton, M., & Wolsky, K. (2022). *The barriers to faculty reporting incidences of academic misconduct at community colleges.* In: S.E. Eaton & J. Christensen Hughes. (Eds.), *Academic integrity in Canada* (Vol 1, pp. 467–485). Springer. https://doi.org/10.1007/978-3-030-83255-1_24

Hogan, B. E., & Trotter, L. D. (2013). Academic freedom in Canadian higher education: Universities, colleges, and institutes were not created equal. *Canadian Journal of Higher Education, 43*(2), 68–84.

International Center for Academic Integrity [ICAI]. (2021). *The fundamental values of academic integrity.* (3rd ed.). https://academicintegrity.org/images/pdfs/20019_ICAI-Fundamental-Values_R12.pdf

Jones, G. A. (2014). Building and strengthening policy research capacity: Key issues in Canadian higher education. *Studies in Higher Education, 39*(8), 1332–1342.

Kenny, N., & Eaton, S. E. (2022). Academic integrity through a SoTL lens and 4M framework: An institutional self-study. In S. E. Eaton & J. Christensen Hughes (Eds.), *Academic integrity in Canada. Ethics and integrity in educational contexts* (Vol. 1, pp. 573–592). Springer. https://doi.org/10.1007/978-3-030-83255-1_30

Kutbi, M., Al-Hoorie, A. H., & Al-Shammari, A. H. (2024). Detecting contract cheating through linguistic fingerprint. *Humanities and Social Sciences Communications, 11*(664). https://doi.org/10.1057/s41599-024-03160-9

McCowan, T. (2015). Should universities promote employability? *Theory and Research in Education, 13*(3), 267–285.

Miron, J., Eaton, S. E., McBreairty, L., & Baig, H. (2021). Academic integrity education across the Canadian higher education landscape. *Journal of Academic Ethics, 19*, 441–454. https://doi.org/10.1007/s10805-021-09412-6

Quality Assurance Agency for Higher Education. (2022). *Contracting to cheat in higher education: How to address essay mills and contract cheating*. https://www.qaa.ac.uk/docs/qaa/guidance/contracting-to-cheat-in-higher-education-2nd-edition.pdf

Reid, J. (2016). Redefining "employability" as something to be achieved: Utilising Tronto's conceptual framework of care to refocus the debate. *Higher Education, Skills and Work-Based Learning, 6*(1), 55–68.

Robinson, J. A., & Glanzer, P. L. (2017). Building a culture of academic integrity: What students perceive and need. *College Student Journal, 51*(2), 209–221. https://link.gale.com/apps/doc/A503260892/AONE?u=anon~a080fe99&sid=googleScholar&xid=8ad26e02

Seeland, J., Cliplef, L., Munn, C., & Dedrick, C. (2022). Mathematics and academic integrity: Institutional support at a Canadian college. *International Journal of Mathematical Education in Science and Technology, 53*(3), 673–680. https://doi.org/10.1080/0020739X.2021.1981472

Seeland, J., Stoesz, B. M., & Vogt, L. (2020). Preventing online shopping for completed assessments: Protecting students by blocking access to contract cheating websites on institutional networks. *Canadian Perspectives on Academic Integrity, 3*(1). https://doi.org/10.11575/cpai.v3i1.70256

Stephens, J. M. (2016). Creating cultures of integrity: Multilevel intervention model for promoting academic honesty. In T. Bretag (Ed.), *Handbook of academic integrity*. Springer.

Stoesz, B. M., Seeland, J., MacLeod, P., Vogt, L., Markovics, L., Denham, T., Gervais, L., & Usick, B. (2020). Creating a collaborative network to promote cultures of academic integrity in Manitoba's post-secondary institutions. *Canadian Perspectives on Academic Integrity, 3*(1), 39–48. https://doi.org/10.11575/cpai.v3i1.69763

Thacker, E. J., & McKenzie, A. (2022). Using quality assurance frameworks to support an institutional culture of academic integrity at Canadian universities. In S. E. Eaton & J. Christensen Hughes (Eds.), *Academic integrity in Canada: An enduring and essential challenge* (pp. 519–534). Springer.

The Colleges Act,. CCSM c C150.1, https://web2.gov.mb.ca/laws/statutes/ccsm/c150-1.php

Whitley, B. E., Jr., & Keith-Spiegel, P. (2002). *Academic dishonesty: An educator's guide*. Lawrence Erlbaum Associates Publishers.

Yin, R. K. (2014). *Case study research: Design and methods* (5th ed.). SAGE.

Caitlin Munn is an Education Quality Assurance Specialist at Assiniboine College. She is interested in leveraging the intersection of quality assurance and academic integrity to support high-quality learning experiences.

Deanna Rexe is Vice President Academic at Assiniboine College, and an Adjunct Professor in the Faculty of Education and affiliated scholar with the Centre for the Study of Educational Leadership and Policy at Simon Fraser University. Her research examines the relationship between decision contexts and higher education policy choices. As a practitioner and academic leader, she is interested in service, policy, and governance interventions to support increased access and success in higher education.

Josh Seeland is the Manager of Library Services at Assiniboine College, a founding member of the Manitoba Academic Integrity Network (MAIN), and Chair of Assiniboine's Academic Integrity Advisory Committee (AIAC). His writing on academic integrity can be found in academic journals, the book "Contract Cheating in Higher Education," and the book "Academic Integrity and the Role of the Academic Library: Institutional Examples and Promising Practices."

Chapter 3
Building an Academic Integrity Plan: Our Journey at a Mid-Sized Polytechnic

Caitlin Ratcliffe, Sona Macnaughton, Teneil Vuori, and Yvonne Phillips

Abstract Academic integrity is an increasingly prevalent concern in the vocational and polytechnic sectors, particularly for small or mid-sized institutions with limited resources and infrastructure. This chapter presents the journey of a team of librarians to build an academic integrity plan at a mid-sized polytechnic institution in Alberta, Canada, from 2019 to 2023. Divided into three phases, the chapter outlines the initiatives undertaken by the librarian team to develop and implement strategies that support academic integrity during a period of change and growth. These efforts include the creation of general resources for learners at the start of their academic studies, as well as specific resources to support learners who have engaged in academic misconduct. The narrative explores the challenges our team faced as well as our next steps, such as promoting these new resources for learners. The chapter concludes by reflecting on this journey and offering recommendations for educators within the vocational and polytechnic sectors as they strive to promote academic integrity amongst a diverse student population.

Keywords Academic integrity · Information literacy · Education programs · Online tutorials · Web-based instruction · Stakeholder engagement

C. Ratcliffe
NorQuest College, Edmonton, AB, Canada
e-mail: caitlin.ratcliffe@norquest.ca

S. Macnaughton (✉) · T. Vuori · Y. Phillips
Red Deer Polytechnic, Red Deer, AB, Canada
e-mail: sona.macnaughton@rdpolytech.ca; teneil.vuori@rdpolytech.ca; yvonne.phillips@rdpolytech.ca

L. Vogt, B. M. Stoesz (eds.), *Academic Integrity in Vocational and Polytechnic Education*, Ethics and Integrity in Educational Contexts 8,
https://doi.org/10.1007/978-3-031-90091-4_3

Building an Academic Integrity Plan: Our Journey at a Mid-Sized Polytechnic

Academic misconduct is an increasingly prevalent concern within the vocational and polytechnic sectors of higher education (Ferguson et al., 2023; Hamilton & Wolsky, 2022). The sudden availability of generative artificial intelligence tools and the increase in students from diverse educational backgrounds have highlighted the need to promote academic integrity with intention (Moya et al., 2024; Steeves, 2024). However, institutions within these sectors—particularly small or mid-sized institutions—face a distinct set of challenges (Bealle, 2017; Hamilton & Wolsky, 2022), including limited resources or institutional infrastructure, such as the lack of a dedicated office or specialist positions (see Clark et al., 2020; Salmon & Tilley, 2021). This chapter will describe the efforts of a team of librarians from Red Deer Polytechnic (RDP) from 2019 to 2023 as we strove to support and promote academic integrity during a period of change and growth. In this chapter, we explore the complexities of developing a plan for strengthening academic integrity in our mid-sized Canadian institution. This narrative is intended to be an authentic portrayal of our work as shaped by our institutional context and, as such, documents both successes and challenges. Educators within the vocational and polytechnic sectors—including instructors, librarians, writing tutors, teaching and learning supports, and administrators—can benefit from the lessons learned as we collectively endeavour to promote academic integrity among future professionals.

RDP is a mid-sized, publicly funded post-secondary institution in Red Deer, Alberta, Canada. In 2022–2023, RDP reported 4135 Full-Load Equivalent (FLE) enrolments (Government of Alberta, 2023a), including approximately 6402 credit and apprenticeship students and 2257 non-credit students (RDP, n.d.-a). RDP offers a wide range of programming, including apprenticeships, certificates, diplomas, micro-credentials, and degrees. Previously known as Red Deer College, RDP transitioned from a comprehensive community college to a polytechnic institution in 2021. Since 2021, RDP's programming has grown considerably, including the addition of five degrees as well as the expansion of existing programs in high-demand areas such as health sciences and business. With the transition to polytechnic status, there has been an increased institutional focus on building capacity for degree programs, and subsequent research and scholarship expectations.

Like many small or mid-sized institutions, RDP has no centralized academic integrity office. The responsibilities associated with academic integrity are shared between instructors, administrators (primarily the Associate Deans), and the Division of Student Success, which includes the Library. The Library's involvement with academic integrity education at RDP is consistent with many publicly-funded institutions of higher education in Canada (Miron et al., 2021). Morrow (2022) argued that academic librarians, through their inherent expertise in information literacy, have a key role in advocating for and supporting students' integrity in academic work. At RDP, the recent initiatives to strengthen academic integrity were driven by a team of librarians in response to concerns from stakeholders across the institution. These initiatives occurred in three phases: laying the foundation for academic integrity, designing supports for academic misconduct, and building proactive academic integrity education.

Phase 1: Laying the Foundation for Academic Integrity

The first phase began in mid-2019 when an instructor, inspired by their experience with a student academic integrity tutorial at the University of Calgary, approached one of the librarians to suggest developing a similar tutorial. This suggestion spurred the librarians to explore options for resources specific to academic integrity that could supplement the existing Library resources on plagiarism and citation. With this goal in mind, the librarian team began preliminary exploration to identify existing resources and best practices. A librarian representative began attending the provincial Alberta Council on Academic Integrity (ACAI) meetings, which continued until the COVID-19 pandemic disruption started in March 2020. Through these meetings, the librarian team learned about how other institutions in Alberta handle academic integrity and what types of tutorials and educational materials were being developed.

In April 2020, the team began working on an Academic Integrity Guide, a website which was built using Springshare's LibGuides platform, and shared with institutional stakeholders in Spring 2020. The guide was targeted towards learners and included basic information about different aspects of academic integrity, including a series of comic-based scenarios originally developed at the University of Waterloo, but it was not a full tutorial. Although it was clear that more resources were still required to support academic integrity at RDP, the foundational phase ended after mid-2020. With the sudden shift to online-only learning and services, maintaining the Library's existing operations became a higher priority than creating new resources.

Phase 2: Designing Supports for Academic Misconduct

The second phase began in early 2022. In this phase, the librarian team focused on developing supports for students who engaged in academic misconduct. This phase began when the policy governing student academic misconduct came due for review. The policy review opened up an opportunity for RDP to consolidate two policies (the *Student Academic Integrity and Academic Misconduct Policy* and the *Student Misconduct: Non-Academic Policy*) into a new policy, *Student Rights and Responsibilities*. The procedural elements of the two former policies were converted into two new procedures: the *Academic Misconduct Procedure* and the *Non-Academic Misconduct Procedure*. When the policy review process began, the Dean of Students, who was a member of the policy review committee and the librarians' direct supervisor, asked the librarian team to investigate the current state of academic integrity at RDP.

The librarian team began this investigation by consulting current and past Associate Deans about their experiences. Since Associate Deans play a critical role in responding to academic misconduct at RDP, they were seen as important stakeholders with whom to start this conversation. Although students, faculty, and frontline staff are also key stakeholders in academic integrity, the team limited initial consultation to the Associate Deans due to time and workload constraints. Through semi-structured conversations with the Associate Deans, several themes emerged around the existing challenges with academic misconduct and what was needed to support students, faculty, and administration.

Identified Challenges & Needs

The librarian team quickly identified that academic misconduct was a significant concern across the institution. As was common at many academic institutions during the COVID-19 pandemic (Comas-Forgas et al., 2021; Maryon et al., 2022; San Jose, 2022), the rapid shift to online learning compounded existing problems with assessments, particularly with honesty during virtual exams and assignments. However, the Associate Deans identified that many instances of academic misconduct at RDP fell within the realm of unintentional or accidental misconduct. In these cases, it was clear that students struggled to connect the theory of citation to their practical work. To address this challenge, stakeholders expressed a need for greater education for students about academic integrity, both proactively as well as in response to accidental instances of misconduct. Several stakeholders stressed the importance of reflection as part of this education, encouraging students to recognize how academic and personal integrity applies to their chosen profession.

The librarian team also identified that the existing policy, *Student Academic Integrity and Academic Misconduct,* created two challenges for faculty and administration. Firstly, the policy took a punitive approach to academic misconduct. While this approach was appropriate for certain instances of deliberate misconduct, it was not suitable for students who unintentionally engaged in academic misconduct due to a lack of knowledge or skills (Bertram Gallant & Drinan, 2008; Bertram Gallant, 2017; Bretag et al., 2011; Bretag et al., 2014; Morris, 2016; Morris & Carroll, 2016). There was no formal process built into the policy or procedure for these students to learn about their errors beyond the varying levels of feedback provided by individual instructors or Associate Deans. Students could accumulate multiple infractions—sometimes within the span of a few days—when assignments were submitted in different courses at the end of the term. The consequences could then be severe, without giving students any opportunity to learn from their errors. Secondly, instructors and administrators reported that the policy was hard to understand and follow. As a result, faculty members were inclined to handle instances of academic misconduct informally on their own rather than navigate the policy requirements. Difficulties understanding the policy, combined with the policy's overly punitive approach, contributed to chronic under-reporting of academic misconduct and inconsistent handling of incidents. Due to the under-reporting and the lack of institution-level tracking, students could commit several "first" offenses without understanding the seriousness of continued unethical practices.

To address the issues with the policy, stakeholders expressed the need for clear procedural directives regarding reporting and tracking, as well as updates to the policy to incorporate educational opportunities for students. Additionally, stakeholders expressed a desire for more training for faculty in identifying academic misconduct and applying the policy and procedures. Stakeholders hoped these changes would lead to a more consistent approach in responding to academic misconduct by individual faculty members and programs.

Creation of Remedial Tutorial & Remediation Process

Based on these conversations with stakeholders, the librarian team began developing a remedial tutorial in April 2022. The remedial tutorial is intended to be assigned following a first offense of academic misconduct. Despite negative connotations associated with the term "remedial," the team selected this nomenclature intentionally as it aligns with the terminology used in other institutional policies and procedures at RDP. To begin, the librarian team conducted an environmental scan of similar tutorials. The team was impressed by the "Understanding Academic Integrity" tutorial from Sheridan College (Sheridan College, 2023), including the visualization of the remediation process (Sheridan College, n.d.). The RDP librarians requested and received permission from Sheridan College's Academic Integrity Office to adapt this tutorial. The RDP remedial tutorial was built using Springshare's LibWizard platform.

The RDP remedial tutorial, as adapted from Sheridan College, consists of 11 multiple-choice questions about nine topics:

1. Assignment collaboration
2. Contract cheating
3. Plagiarism
4. Incorrect citations
5. Missing citations
6. Self-plagiarism
7. Sharing text/exam answers
8. Using course sharing sites
9. Using homework help sites

Each topic includes a short video scenario, a written explanation of the error in the video, and best practices to avoid the error, including recommendations to RDP services such as the Writing Tutor. In addition to this content from Sheridan College, the librarian team added two reflection questions based on suggestions from the RDP stakeholders:

1. Based on this tutorial, consider your case of academic misconduct. What happened? Discuss what you will do differently next time.
2. What is your intended career? Discuss the connection between academic integrity and professional conduct.

Students must receive 100% on the multiple-choice section before they advance to the reflection questions. Inspired by the process at Sheridan College, the librarian team identified the need for an academic integrity specialist during the environmental scan phase. The addition of a specialist was an important element of the new process, as dedicated specialists can play a vital role in promoting a culture of academic integrity (Bertram Gallant & Drinan, 2008; Morris & Carroll, 2016; Vogt & Eaton, 2022). In the context of the remedial process at RDP, the librarian team wanted the specialist to review and discuss the tutorial results (particularly the reflection questions) one-on-one with individual students. To this end, the Library administration identified that the Academic Support Coordinator—the staff member who managed the Library's tutoring services—would meet with these students to complete the remedial process. The Library administration selected this staff member due to their background in English language writing skills. Although the staff member's pre-existing managerial responsibilities did not change, the position's title was changed to Academic Integrity Coordinator to reflect the addition of these responsibilities and to make this individual a more easily identifiable part of the new process. Despite the new title, this staff member was not involved with the remedial tutorial creation process nor the subsequent development of further academic integrity-related resources. In the future, this position may be expanded to include promotion and periodic review of the remedial process, as well as providing input into the revision and creation of resources.

As the second phase was being finalized, a cohort-based program was selected to pilot the proposed remedial process prior to its inclusion in policy and procedure. In December 2022, 14 students were identified for the pilot; each student was nearing completion of their first semester and had an academic misconduct offense due to a lack of knowledge or skills. Based on feedback from the pilot, the librarian team made minor adjustments to the tutorial and the overall process, such as increasing communication between the Academic Integrity Coordinator and the course instructors. At the conclusion of the pilot, the librarian team promoted the new remedial resources to the Associate Deans. The team also provided feedback to the policy review sub-committee with suggestions on how to integrate the remedial resources into the procedure documentation. Following the pilot, only two additional students completed the remedial process by December 2023, with a few more who completed the tutorial but did not meet with the Academic Integrity Coordinator. The limited uptake of the remedial process will be discussed in the *Future Directions* section of this chapter.

Phase 3: Building Proactive Academic Integrity Education

Although the librarian team had been developing resources to support proactive academic integrity education sporadically since 2020, the third phase began in earnest in late 2022. During this phase, the librarian team focused on creating new educational resources and opportunities to help students proactively avoid academic misconduct.

Collaboration with RDP International Education

One aspect of the third phase included collaboration with RDP International Education. In the past decade, the number of international students attending RDP has increased a staggering 933%, from 75 unique students (1.1% of total headcount) in 2012–2013 to 775 unique students (12.1% of total headcount) in 2022–2023 (Government of Alberta, 2023b). In this rapidly changing landscape, stereotypes emerged about the prevalence of international students breaching academic integrity; however, studies examining the rates of academic misconduct based on student ethnicity or visa status have demonstrated that these identifying factors are not predictors of academic integrity breaches (Fass-Holmes, 2017; Martin et al., 2011). International students enrolled in Western institutions report that expectations regarding academic integrity can differ between various countries and specific academic settings (Christoph, 2016; Click, 2016; Torres-Arends, 2023), particularly in relation to originality (Hayes & Introna, 2005; James et al., 2019; Kaur, 2019; Szilagyi, 2014) and duplicate submission/self-plagiarism (Sanni-Anibire et al., 2021). As a result, culturally aware education about academic integrity is essential for international students to understand the expectations of the institution and the requirements of their specific courses (Fass-Holmes, 2017; Martin et al., 2011).

To better support the needs of the growing international student population, in March 2022, the librarian team created and contributed content to RDP International Education's mandatory pre-arrival orientation for new international students, which consists of asynchronous resources and activities, synchronous online seminars, and a two-day in-person event at the start of each semester. The asynchronous modules include learning outcomes related to academic integrity, an interactive video using Camtasia/Screencast, and a corresponding academic integrity quiz in RDP's learning management system, Blackboard. For the in-person orientation events, the librarians teach a 30–45-minute scenario-based workshop at the start of every semester. Although the content is a high-level overview, these introductory resources ensure that RDP takes responsibility for communicating academic integrity expectations to new international students entering the Canadian post-secondary context.

Creation of Fundamentals Tutorial

The third phase also included the development of a fundamentals tutorial. Unlike the remedial tutorial, which was designed to support learners who engaged in a first offense of academic misconduct, the fundamentals tutorial was designed to educate all learners at the start of their academic journey. The "fundamentals" nomenclature was intentionally selected to convey that this tutorial builds a practical foundation of knowledge about academic integrity; this nomenclature also parallels RDP Library's citation fundamentals tutorials. The team conducted an environmental scan in October 2022, focusing on tutorials from the United States and Canada with interactive components, such as quizzes and videos, that were openly licensed to enable adaption and re-distribution.

After identifying tutorials that met the initial criteria, the team further refined our selection to focus on those that attended to the diversity of learners at RDP. The team prioritized resources containing strong elements of equity and inclusion to reflect the core values of librarianship (American Library Association, 2024) as well as RDP's institutional values (RDP, n.d.-b). To this end, the team examined content through the lens of cultural humility, predicated on the understanding that concepts of academic integrity can be culturally dependent. While there is limited literature exploring the relationship between cultural humility and academic integrity, cultural humility—an ongoing process of self-reflection that starts with recognizing how our own cultural identities influence how we interact with others (Yeager & Bauer-Wu, 2013)—is a foundational component of providing library services to a diverse range of patrons (Hurley et al., 2019; Hurley et al., 2022). In addition, the team recognized the importance of creating accessible educational materials, given the many inequitable and ableist academic integrity practices and policies at post-secondary institutions (Gagné, 2024) and the need for academic integrity educators to create ethical and accessible resources (Pagaling et al., 2022). When reviewing existing content, the team emphasized the principles of Universal Design for Learning and strove to include multiple means of engagement, representation, and expression throughout the tutorial. The final selection process was also influenced by the team's awareness of our own limitations, such as the lack of time and specialized resources to create videos.

The team ultimately selected a tutorial from Simon Fraser University (SFU, 2023) as the base due to its flexible modular approach and the use of the first-person voice, which implies both agency and ownership in learning about academic integrity. The team supplemented the SFU tutorial with content from Northern Illinois University (n.d.), the University of Waterloo (n.d.), and Assiniboine Community College (n.d.). It was vital to track the source of different elements to acknowledge our use and modification of the content; therefore, the team used extensive outlining and shared document collaboration throughout the creative process.

The RDP fundamentals tutorial includes five topics:

1. Introduction to Academic Integrity
2. Plagiarism

3. Inappropriate collaboration
4. Cheating (which is further subdivided into Exam Cheating and Contract Cheating)
5. Misrepresentation.

Each topic includes a written introduction to the concept, "real life" video and comic-based scenarios, practical tips and strategies, and a non-graded quiz with immediate feedback to check understanding. With the emergence of mainstream generative artificial intelligence in late 2022, the RDP team also developed original content for the plagiarism topic regarding the ethical use of tools such as ChatGPT. The team created a character, *Academic Integrity Alex* (Fig. 3.1), to add a human element to navigating the information. This character was intentionally designed as non-binary with multiple skin tones to make a relatable persona for a diverse range of students.

The RDP fundamentals tutorial (Fig. 3.2) was built using Springshare's LibGuides and LibWizard platforms. The RDP fundamentals tutorial is licensed as Creative Commons Attribution-NonCommercial-ShareAlike 4.0 with only a few exceptions, as noted within the tutorial.

Referrals to relevant RDP services, such as the Writing Tutors, are incorporated throughout. At the culmination of the tutorial, successful completion of a graded final quiz generates a "Certificate of Completion"with the participant's name and grade. Participants have unlimited attempts to complete the final quiz, but the certificate is only generated if participants achieve 80% or higher (12 or 15 questions correct). Instructors may direct students to specific topics within the tutorial or require them to complete it and submit their PDF certificate.

The fundamentals tutorial was published and shared widely with RDP faculty in September 2023. The Dean of Students and the Associate Dean of the Library highlighted the tutorial in their presentation about the new policy and procedures and shared it with every School (the highest level of academic administration in the organizational structure at RDP) throughout Fall 2023. From September to December 2023, the final quiz of the fundamentals tutorial was completed 227 times by 188 unique students; in Winter 2024, the final quiz was completed 723

Fig. 3.1 *Academic Integrity Alex. Note.* Screenshot from Red Deer Polytechnic Library. (2023). *Academic integrity fundamentals tutorial.* https://guides.rdpolytech.ca/academicintegrityfundamentals. Academic Integrity Alex is adapted from Adobe Stock (Education License) and may not be re-used or modified. This image has been converted to grayscale for the purposes of publication; to view the character in full colour, please visit the Red Deer Polytechnic Library *Academic Integrity Fundamentals Tutorial*

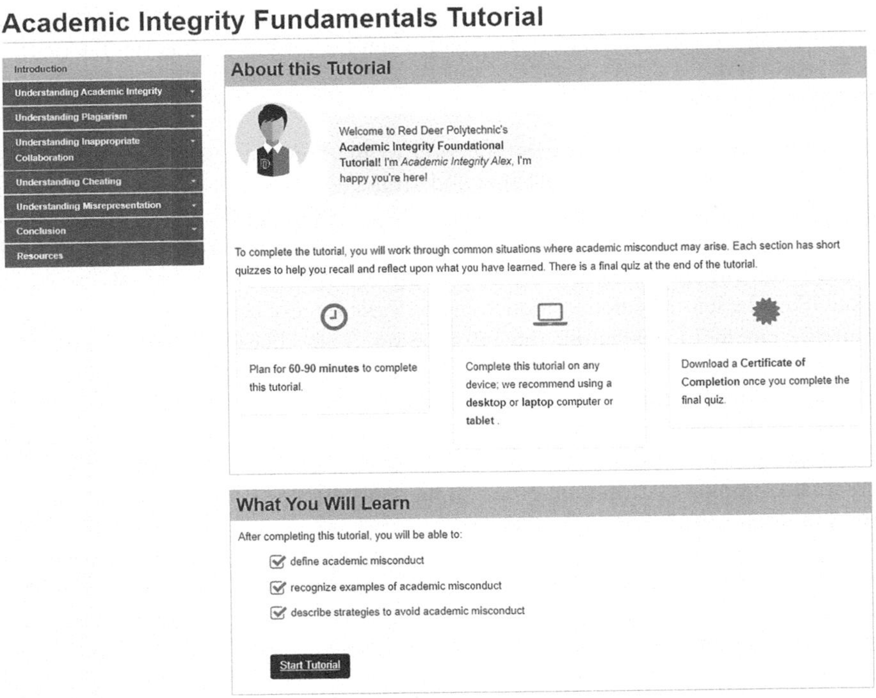

Fig. 3.2 *Academic Integrity Fundamentals Tutorial. Note.* Screenshot from Red Deer Polytechnic Library. (2023). *Academic integrity fundamentals tutorial.* https://guides.rdpolytech.ca/academicintegrityfundamentals

times by 296 unique students. The librarian team anticipates further uptake in the 2024–2025 academic year as awareness of the tutorial increases.

Future Direction

With the creation of these new resources to both proactively educate learners as well as respond to academic misconduct, the librarian team is looking to the future. In addition to periodically reviewing and updating the new resources as the academic integrity landscape changes, the team's next priorities involve advocating for the remedial process, promoting the fundamentals tutorial, and creating educational opportunities for faculty.

Advocating for the Remedial Tutorial & Process

To continue supporting students who have engaged in academic misconduct, the librarian team's next steps will be advocating for the new remedial process. This advocacy will focus on addressing three aspects of the new remedial process. The first is increasing uptake. During the first year following the launch of the remedial process, only two students completed the entire process (excluding the 14 students who participated in the pilot), with a few more completing the tutorial but not meeting with the Academic Integrity Coordinator. The librarian team reached out about this limited uptake, and stakeholders indicated they would be more likely to use the remedial process if it were explicitly named in the *Academic Misconduct Procedure.* Initially, the procedure identified "attendance at academic integrity seminars or completion of similar educational and well-being activities" as one possible consequence for a first incident of academic misconduct but did not explicitly reference the Library's remedial tutorial or process (RDP, 2023, p. 3). Based on the feedback from stakeholders, the librarian team advocated for the remedial resources to be named in the institutional procedure. The subsequent version of the procedure now also explicitly states that the instructor may provide "direction, when applicable, to contact the responsible Polytechnic employee to access the Academic Integrity remedial course and to set up an appointment to review the results" (RDP, 2023, p. 4), although it is not mandatory. The librarian team will continue promoting these resources across the institution to improve uptake.

If uptake of the remedial process increases, a potential concern is workload. Schneider (2023) argued that upholding a culture of academic integrity, especially when responding to academic misconduct, is a workload issue. At RDP, it is unclear how further uptake of the remedial process will impact the Academic Integrity Coordinator, who presently serves as an impartial third party to discuss the remedial tutorial and refer students to other academic supports. The academic integrity-related responsibilities and the title of the Academic Integrity Coordinator are new additions to an existing position at RDP, and it is possible that the responsibilities of this position may change over time. It is also unknown whether the additional work related to academic integrity will be manageable for one position with existing responsibilities. If the workload of the Academic Integrity Coordinator becomes unsustainable, the Library's administration is committed to addressing this issue. If necessary, a pyramid framework, such as the one proposed by Ellis and Murdoch (2024), could be used to prioritize and operationalize responses to academic misconduct based on risk and cost, ensuring the scalability of the remedial process at RDP.

The final aspect is tracking the completion of the remedial process. Currently, instructors individually refer students to the remedial tutorial and the Academic Integrity Coordinator, who reports back to the instructor when the student has successfully completed the tutorial and the meeting. Initially, academic misconduct offenses were tracked by the Office of the Registrar. This work has recently shifted to the Office of the Dean of Students (which includes the Library), although the institutional procedures for reporting and tracking academic misconduct have not yet been formalized. Tracking completion of the remedial process is important; if a subsequent instance of academic misconduct occurs, students cannot thus plead ignorance of the tenets of academic integrity nor of the resources available. The librarian team will continue working with the Academic Integrity Coordinator and other internal stakeholders to integrate the remedial process into institutional practices.

Promoting the Fundamentals Tutorial

While the fundamentals tutorial currently has more institutional awareness than the remedial process, the librarian team is also working to promote this foundational resource further. Integration into a first-year course would be ideal; however, there is no suitable course required by all or even most programs at RDP. To reach more students, the librarian team is pursuing options to incorporate the fundamentals tutorial more systematically throughout the institution's learning management system, increasing the chances that more students will find the tutorial even if it is not assigned by instructors.

Creating Education Opportunities for Faculty

The librarian team has also identified the need for further faculty education to help instructors promote academic integrity, identify and investigate academic misconduct, and effectively utilize the institutional policy and procedure to respond to academic misconduct. In the context of colleges and technical institutes, Schneider (2023) argued that "faculty, regardless of employment status or experience level, should have access to professional development opportunities specific to the unique challenges of identifying, investigating, and reporting academic integrity violations" (pp. 9–10). Similarly, faculty can benefit from support around creating "learning-oriented" environments that reduce the possibility of, and learners' inclination towards, academic misconduct (Bertram Gallant, 2008, 2017). When coming directly from their professional context into a post-secondary environment, faculty may feel disconnected from academic practices, such as identifying and responding to academic misconduct (Hamilton & Wolsky, 2022). Additionally, faculty may be unfamiliar with institution-specific policies, procedures, and available resources. These feelings of discomfort can be compounded by the pressure on instructors and students to adapt to the rapidly changing environment of post-secondary education in Canada, including increased student-to-instructor ratios as well as the dynamic evolution of tools with the potential for misuse, such as generative artificial intelligence. The librarian team at RDP is investigating options for education, such as an openly licensed asynchronous tutorial for faculty, to address these needs.

Recommendations

Based on our experience developing a plan for strengthening academic integrity, we have four recommendations for educators in the vocational and polytechnic sectors:

Assess Your Situation

Prior to making changes, it is important to assess your institutional context. The McCabe-ICAI Academic Integrity Surveys (International Center for Academic Integrity, n.d.), the Academic Integrity Rating System (Bertram Gallant, 2022), and the Academic Integrity Maturity Model (Glendinning, 2014) are useful tools for assessing your institutional context. In general, questions to consider may include:

- What does academic misconduct look like for your learners? What problems are occurring?
- Does your institution have data regarding misconduct that can be used to corroborate the need for more support and resources?
- How is academic misconduct handled at your institution? What is your current policy?
- Is there a role or a unit that spearheads the efforts to promote academic integrity on campus and build a community to support institutional members?
- What do stakeholders, such as faculty members or relevant administration, at your institution think?
- When considering your institution's approach to academic integrity, what aspects are working well? Where are opportunities for change?

The answers to these questions should shape the direction of your work. In some cases, only small changes to existing processes may be needed; in others, the system may instead need an overhaul requiring new policies, procedures, and/or resources. By connecting with stakeholders in the initial stages, you can identify the top priorities in your institutional context. These conversations also create greater buy-in from stakeholders when new resources or procedures are implemented.

Identify Existing Resources and Capacity

Next, it is important to identify the existing resources available at your institution to support your goals. If your institution already has a department spearheading academic integrity initiatives, your focus may be on promoting these existing services and resources. Alternatively, you may have individual personnel, such as librarians or writing tutors, who provide academic integrity-related services and have the capacity to develop new resources.

In addition, it is valuable to identify existing resources available at other institutions. Many academic integrity-related resources (such as ours) are openly licensed with Creative Commons licenses that permit re-use and adaptation. Even without a license granting explicit permission, many institutions are willing to provide permission for others to adapt and re-use their resources. Conducting an environmental scan early in your process can save you the time and effort of creating wholly new resources.

Once you have identified the available resources, you can refer back to your goals and identified problems to create a plan. It is helpful to determine which goals can be addressed immediately and which may need to wait for the future, as some projects may require additional funding, staff, or resources to implement.

Consider Different Perspectives

As you plan and develop your resources, it is valuable to consider best practices for equity and inclusion. Eaton (2024) noted that "there is no single universal understanding of academic integrity that fits every single context. Instead, there are multiple and complementary understandings" (p. 19). Educators should seek to comprehend the nuances between individuals' varying conceptions of academic integrity and "resist the temptation to conflate difference with deficiency" (Eaton, 2024, p. 19). In our work, we strove to use the practice of cultural humility to acknowledge the influence of individual and cultural positionality on understandings of academic integrity. Embracing multiple, complementary understandings of academic integrity represents a shift away from current practices and policies, which Gagné (2024) argued are largely inequitable and ableist due to inaccessible assessment design and pedagogies. Following Gagné's suggestion to improve accessibility, we integrated the principles of Universal Design for Learning, including multiple means of engagement, representation, and expression, when creating new tutorials. For example, we added multiple means of representation by enhancing the traditional textual content with videos (including transcriptions), comic-based scenarios, and non-graded interactive quizzes. Educators should seek to prioritize diverse perspectives throughout their efforts to promote academic integrity.

Inviting feedback from individuals less familiar with the project can also offer fresh perspectives. Stakeholders such as instructors, students, writing tutors, library staff, and staff supporting international students can provide valuable feedback. This external review can help identify any overlooked areas and ensure that the tone is inclusive and supportive to learners from diverse backgrounds.

Integrate with Policy and Procedure

Throughout our journey, the importance of institutional policy and procedure became very clear. This observation is echoed by Morris and Carroll (2016), who highlight that regular review and consistent implementation of academic integrity-related policies facilitate institutional change. From the perspective of on-the-ground practitioners, including specific processes or resources in policy and procedure can positively impact the implementation of new initiatives. In our case, stakeholder feedback indicated that there would be greater uptake of the remedial process if it were explicitly mentioned in the new *Academic Misconduct Procedure*. To promote consistent uptake of new initiatives across the institution, it is ideal to draw upon policy and procedure whenever possible and to review these documents regularly to ensure they are relevant and comprehensible to stakeholders (Morris & Carroll, 2016; Shane et al., 2018).

Grow as You Are Able

Depending on the resources available to you, it may not be possible to implement all of your desired goals at once. For example, we developed our remedial process before creating our fundamentals tutorial, even though stakeholders expressed a clear desire for both resources. The time investment for initiatives such as these can vary considerably. At RDP, four librarians contributed to these initiatives, in addition to our existing responsibilities, over a five-year period, which included the pandemic disruption that started in March 2020. At other vocational or polytechnic institutions, similar initiatives may take more or less time depending on the number of people included in the project, the amount of time they can commit, and existing expertise. Morris and Carroll (2016) noted that holistic institutional approaches to academic integrity can take 3 to 5 years before positive results may be observed. Though academic integrity may seem urgent, pacing yourself is key when implementing sustainable initiatives.

Conclusion

Overall, the RDP librarian team has found supporting academic integrity to be a rewarding process. Although our work has been imperfect at times, we are proud to share our ongoing journey for strengthening academic integrity through a period of change and growth. We hope this narrative will benefit educators within vocational and polytechnic institutions as we collectively continue to promote academic integrity as a vital component of our learners' scholarly and professional lives.

References

American Librarian Association. (2024, January). *Core values of librarianship*. https://www.ala.org/advocacy/advocacy/intfreedom/corevalues.

Assiniboine Community College. (n.d.). *Academic integrity for students: Misrepresentation*. https://assiniboine.libguides.com/ai-students/misrepresentation

Bealle, P. (2017). Community college academic integrity lessons that put research into practice. *Theory Into Practice, 56*(2), 144–455. https://doi.org/10.1080/00405841.2017.1283573

Bertram Gallant, T. (2008). *Academic integrity in the twenty-first century: A teaching and learning imperative*. Jossey-Bass.

Bertram Gallant, T. (2017). Academic integrity as a teaching and learning issue: From theory to practice. *Theory Into Practice, 56*(2), 88–94. https://doi.org/10.1080/00405841.2017.1308173

Bertram Gallant, T. (2022). *'When a thousand flowers bloom': The academic integrity rating system (AIRS)*. International Center for Academic Integrity. https://academicintegrity.org/resources/blog/96-2022/february-2022/345-when-a-thousand-flowers-bloom-the-academic-integrity-rating-system-airs.

Bertram Gallant, T., & Drinan, P. (2008). Toward a model of academic integrity institutionalization: Informing practice in postsecondary education. *Canadian Journal of Higher Education, 38*(2), 24–43. https://doi.org/10.47678/cjhe.v38i2.508

Bretag, T., Mahmud, S., Wallace, M., Walker, R., James, C., Green, M., East, J., McGowan, U., & Partridge, L. (2011). Core elements of exemplary academic integrity policy in Australian higher education. *International Journal for Educational Integrity, 7*(2), 3–12. https://doi.org/10.21913/IJEI.v7i2.759

Bretag, T., Mahmud, S., Wallace, M., Walker, R., McGowan, U., East, J., Green, M., Partridge, L., & James, C. (2014). 'Teach us how to do it properly!' An Australian academic integrity student survey. *Studies in Higher Education, 39*(7), 1150–1169. https://doi.org/10.1080/03075079.2013.777406

Christoph, M. (2016). *International undergraduate students' perspectives on academic integrity: A phenomenological approach* [Master's thesis, University of Manitoba]. MSpace. http://hdl.handle.net/1993/31757

Clark, A., Goodfellow, J., & Shoufani, S. (2020). Examining academic integrity using course-level learning outcomes. The Canadian Journal for the Scholarship of Teaching and Learning, *11*(2), Article 7. https://doi.org/10.5206/cjsotl-rcacea.2020.2.8508

Click, A. B. (2016). *Adapting to higher education in a new culture: International students' perspectives on research, writing, and academic integrity* [Doctoral dissertation, University of North Carolina]. Carolina Digital Repository. https://cdr.lib.unc.edu/concern/dissertations/6h440t45h

Comas-Forgas, R., Lancaster, T., Calvo-Sastre, A., Sureda-Negre, J. (2021). Exam cheating and academic integrity breaches during the COVID-19 pandemic: An analysis of internet search activity in Spain. *Heliyon, 7,* Article e08233. https://doi.org/10.1016/j.heliyon.2021.e08233

Eaton, S. E. (2024). Global perspectives on academic integrity: Introduction. In S. E. Eaton (Ed.), *Second handbook of academic integrity* (pp. 17–22). Springer. https://doi.org/10.1007/978-3-031-54144-5_134

Ellis, C., & Murdoch, K. (2024). The educational integrity enforcement pyramid: A new framework for challenging and responding to student cheating. *Assessment & Evaluation in Higher Education*. https://doi.org/10.1080/02602938.2024.2329167

Fass-Holmes, B. (2017). International students reported for academic integrity violations: Demographics, retention, and graduation. *Journal of International Students, 7*(3), 644–669. https://doi.org/10.32674/jis.v7i3.292

Ferguson, C. D., Toye, M. A., & Eaton, S. E. (2023). Contract cheating and student stress: Insights from a Canadian community college. *Journal of Academic Ethics, 21*, 685–717. https://doi.org/10.1007/s10805-023-09476-6

Gagné, A. (2024). Academic integrity, ableist assessment design, and pedagogies of disclosure. In S. E. Eaton (Ed.), *Second handbook of academic integrity* (pp. 1245–1260). Springer. https://doi.org/10.1007/978-3-031-54144-5_134

Glendinning, I. (2014). *Assessing maturity of institutional policies for underpinning academic integrity*. International Integrity & Plagiarism Conference Proceedings. https://www.plagiarism.org/paper/assessing-maturity-of-institutional-policies

Government of Alberta. (2023a, February 13). *Full Load Equivalent (FLE) enrolment within the Alberta post-secondary education system*. https://open.alberta.ca/opendata/system-wide-fle-enrolment-within-the-alberta-post-secondary-education-system

Government of Alberta. (2023b, February 13). *Headcount enrolment within the Alberta post-secondary education system*. https://open.alberta.ca/opendata/system-wide-headcount-enrolment-within-the-alberta-post-secondary-education-system.

Hamilton, M., & Wolsky, K. (2022). The barriers to faculty reporting incidences of academic misconduct at community colleges. In S. E. Eaton & J. Christensen Hughes (Eds.), *Academic integrity in Canada: An enduring and essential challenge* (pp. 467–485). Springer. https://doi.org/10.1007/978-3-030-83255-1_24

Hayes, N., & Introna, L. D. (2005). Cultural values, plagiarism, and fairness: When plagiarism gets in the way of learning. *Ethics & Behavior, 15*(3), 213–231. https://doi.org/10.1207/s15327019eb1503_2

Hurley, D. A., Kostelecky, S. R., & Townsend, L. (2019). Cultural humility in libraries. *Reference Services Review, 47*(4), 544–555. https://doi.org/10.1108/RSR-06-2019-0042

Hurley, D. A., Kostelecky, S. R., & Townsend, L. (2022). *Cultural humility*. ALA Editions.

International Center for Academic Integrity. (n.d.). *McCabe-ICAI academic integrity surveys*. https://academicintegrity.org/programs/mccabe-icai-academic-integrity-survey

James, M. X., Miller, G. J., & Wyckoff, T. W. (2019). Comprehending the cultural causes of English writing plagiarism in Chinese students at a western-style university. *Journal of Business Ethics, 154*(3), 631–642. https://doi.org/10.1007/s10551-017-3441-6

Kaur, D. (2019). Academic experiences of Indian international students in US higher education. *American Journal of Creative Education, 2*(1), 1–12. https://doi.org/10.20448/815.21.1.12

Martin, D., Rao, A., & Sloan, L. (2011). Ethnicity, acculturation, and plagiarism: A criterion study of unethical academic conduct. *Human Organization, 70*(1), 88–96. https://doi.org/10.17730/humo.70.1.nl775v2u633678k6

Maryon, T., Dubre, V., Elliott, K., Escareno, J., Fagan, M. H., Standridge, E., & Lieneck, C. (2022). COVID-19 academic integrity violations and trends: A rapid review. *Education Sciences, 12*(12), 901. https://doi.org/10.3390/educsci12120901

Miron, J., Eaton, S. E., McBreairty, L., & Baig, H. (2021). Academic integrity education across the Canadian higher education landscape. *Journal of Academic Ethics, 19*, 441–454. https://doi.org/10.1007/s10805-021-09412-6

Morris, E. J. (2016). Academic integrity: A teaching and learning approach. In T. Bretag (Ed.), *Handbook of academic integrity* (pp. 1037–1054). Springer. https://doi.org/10.1007/978-981-287-098-8

Morris, E. J., & Carroll, J. (2016). Developing a sustainable holistic institutional approach: Dealing with realities 'on the ground' when implementing an academic integrity policy. In T. Bretag (Ed.), *Handbook of academic integrity* (pp. 449–462). Springer. https://doi.org/10.1007/978-981-287-098-8

Morrow, L. (2022). Beyond the traditional: Academic integrity in Canadian librarianship. In S. E. Eaton & J. Christensen Hughes (Eds.), *Academic integrity in Canada: An enduring and essential challenge* (pp. 449–464). Springer. https://doi.org/10.1007/978-3-030-83255-1_23

Moya, B., Eaton, S. E., Pethrick, H., Hayden, A., Brennan, R., Wiens, J., & McDermott, B. (2024). Academic integrity and artificial intelligence in higher education (HE) contexts: A rapid scoping review. *Canadian Perspectives on Academic Integrity, 7*(3). https://doi.org/10.55016/ojs/cpai.v7i3.78123

Northern Illinois University. (n.d.). *Academic integrity: Student tutorial.* https://www.niu.edu/academic-integrity/students/index.shtml

Pagaling, R., Eaton, S. E., & McDermott, B. (2022). *Academic integrity: Considerations for accessibility, equity, and inclusion*. University of Calgary PRISM Institutional Repository. http://hdl.handle.net/1880/114519.

Red Deer Polytechnic. (2023, August 21). *Academic misconduct procedure.* https://rdpolytech.ca/sites/default/files/2023-12/Academic%20Misconduct%20Procedure.pdf.

Red Deer Polytechnic. (n.d.-a). *Red Deer Polytechnic Annual Report 2022/2023.* https://rdpolytech.ca/sites/default/files/2023-12/RDP%20Annual%20Report%20and%20Financial%20Statements%202022%202023_web.pdf

Red Deer Polytechnic. (n.d.-b). *Mission, vision & values.* https://rdpolytech.ca/about-us/leadership-governance/mission-vision-values

Salmon, H., & Tilley, J. (2021). Scaffolding the learning opportunities: Academic integrity at Douglas College. *Canadian Perspectives on Academic Integrity, 4*(2), 72–73. https://doi.org/10.11575/cpai.v4i2.74227

San Jose, A. E. (2022). Academic integrity of students during the COVID-19 pandemic: A mixed methods analysis. *European Journal of Education and Pedagogy, 3*(4), 97–103. https://doi.org/10.24018/ejedu.2022.3.4.400

Sanni-Anibire, H., Stoesz, B. M., Gervais, L., & Vogt, L. (2021). International students' knowledge and emotions related to academic integrity at Canadian postsecondary institutions. *International Journal for Educational Integrity, 17*, Article 21. https://doi.org/10.1007/s40979-021-00088-4

Schneider, B. (2023). *ACIFA's position on academic integrity.* Alberta Colleges and Institutes Faculties Association.. https://www.acifa.ca/_files/ugd/16fcf0_c98c6fb9f9b-c4e4880a9f875d754eb4e.pdf.

Shane, M. J., Carson, L., & Edwards, M. (2018). A case study in updating academic integrity policy and procedures. *New Directions for Community Colleges, 2018*(183), 83–93. https://doi.org/10.1002/cc.20320

Sheridan College. (2023, August 31). *Understanding academic integrity.* https://sheridancollege.libguides.com/c.php?g=709220&p=5052694

Sheridan College. (n.d.). *Academic integrity remediation: The process.* https://sheridancollege.libguides.com/ld.php?content_id=35531180

Simon Fraser University. (2023, September 6). *Academic integrity tutorial.* https://www.lib.sfu.ca/help/academic-integrity/academic-integrity-tutorial

Steeves, M. (2024). Addressing equity in academic integrity. *Canadian Perspectives on Academic Integrity, 7*(2) https://journalhosting.ucalgary.ca/index.php/ai/article/view/77686

Szilagyi, A. (2014). Nigerian students' perceptions and cultural meaning construction regarding academic integrity in the online international classroom. *European Journal of Open, Distance and E-Learning, 17*(1), 172–189. https://doi.org/10.2478/eurodl-2014-0012

Torres-Arends, I. (2023). "I thought we were friends": International students challenges in navigating basic academic regulations at a private Canadian university. *Journal of Comparative & International Higher Education, 15*(1), 93–110. https://www.ojed.org/index.php/jcihe/article/view/4436

University of Waterloo. (n.d.). *Undergraduate academic integrity module.* https://uwaterloo.ca/library/research-supports/academic-integrity/undergraduate-academic-integrity-module

Vogt, L., & Eaton, S. E. (2022). Make it someone's job: Documenting the growth of the academic integrity profession through a collection of position postings. *Canadian Perspectives on Academic Integrity, 5*(1), 21–27. https://doi.org/10.11575/cpai.v5i1.74195

Yeager, K. A., & Bauer-Wu, S. (2013). Cultural humility: Essential foundation for clinical researchers. *Applied Nursing Research, 26*(4), 251–256. https://doi.org/10.1016/j.apnr.2013.06.008

Caitlin Ratcliffe MLIS, is the Digital Literacy Librarian at NorQuest College. From 2019–2024, she was a Librarian at Red Deer Polytechnic. She received her Master of Library and Information Studies (MLIS) from the University of Alberta and her Bachelor of Arts (BA) in English and History from the University of Lethbridge. Her work focuses on the experience of students from equity-deserving communities, particularly with regard to academic integrity, and promoting Open Educational Resources to foster equitable learning environments. Her current research examines international students' experiences with library services and of the library as a Place.

Sona Macnaughton MLIS, is a Librarian at Red Deer Polytechnic. She received her Master of Library and Information Studies (MLIS) and Bachelor of Education (BEd) from the University of Alberta. Her varied professional interests include the intersection of artificial intelligence and academic integrity and enhancing user experience in learning spaces. Through her work advocating for open educational resources (OER), she has conducted research to assess faculty and student awareness, use, and perceptions of OER and has led an institution-wide Zero Textbook Cost (ZTC) initiative. She finds joy in data visualization and maker culture.

Teneil Vuori MLIS, is a Librarian at Red Deer Polytechnic. She received her Master of Library and Information Studies (MLIS) and Bachelor of Arts (BA) in Psychology from the University of Alberta. Her professional interests include collection management, information literacy, and learner engagement. As part of her role, she teaches information literacy classes and helps develop resources on citation, plagiarism, and other topics in the realm of academic integrity.

Yvonne Phillips MLIS, is a Librarian at Red Deer Polytechnic. She received her Master of Library and Information Studies (MLIS) from the University of Alberta and her Bachelor of Arts (concentrations in English and Human Geography) from the University of British Columbia. In addition to her librarian interests in fostering positive student experiences and innovating in the library space, she also enjoys connecting with people and participating in the larger RDP community via committee and service work.

Chapter 4
Shifting Perspectives: Academic Integrity in Vocational Education and Training (VET) in the Age of Generative Artificial Intelligence

Nadia Koren

Abstract The term 'academic integrity', typically associated with Higher Education in Australia due to its regulatory requisites imposed by the Tertiary Education Quality and Standards Authority (TEQSA), has an often-overlooked relevance in the Vocational Education and Training (VET) context. Governed by the Australian Skills Quality Authority (ASQA), the VET sector requires evidence of 'authenticity' in all assessments, effectively calling for standards of integrity similar to those of Higher Education. The word 'academic' is often perceived as being reserved for Higher Education, therefore, a cultural shift was required to fully embrace academic integrity in a context where skill demonstrations and apprenticeships are prevalent. This chapter details Technical And Further Education New South Wales (TAFE NSW) proactive approach to embedding academic integrity, addressing misconceptions, and building a shared culture of academic integrity across the institution. Organisational change through collaborative, boundary-spanning techniques forms the basis for academic integrity strategy. The emergence of generative artificial intelligence (genAI) technologies was pivotal to enhancing stakeholder engagement with academic integrity. By outlining the strategies and adaptations specific to the VET context, this narrative aims to inspire other VET institutions and polytechnics to strengthen their academic integrity frameworks, ensuring alignment with local regulations and the evolving technological environment.

Keywords Authenticity · Generative artificial intelligence · Australia · Vocational education and training · Organisational change management · Educational perspective change

---

N. Koren (✉)
TAFE NSW, St Leonards, NSW, Australia
e-mail: nadia.koren1@tafensw.edu.au

L. Vogt, B. M. Stoesz (eds.), *Academic Integrity in Vocational and Polytechnic Education*, Ethics and Integrity in Educational Contexts 8,
https://doi.org/10.1007/978-3-031-90091-4_4

In the Australian context, academic integrity is sometimes perceived as applicable only to higher-level courses, possibly due to the more frequent use of the word 'academic' in the latter context. There is also an assumption that it is difficult for students in vocational education and training (VET) to breach academic integrity expectations because they must demonstrate their practical skills to be eligible for graduation. This assumption, however, ignores the fact that many VET courses do not require practical skills demonstrations and not all competencies can be assessed in this way. Academic integrity is essential in any profession where incompetence could endanger lives—not just in higher education fields like medicine, dentistry, engineering, and nursing. VET teachers and administrators of courses such as electrical trades, health care, early childhood education, construction, and other safety-critical professions must ensure that students are competent when they graduate.

This chapter begins with an overview of the Australian Tertiary Education system, including differences between VET and Higher Education courses and approaches to academic integrity promotion. The argument is also made that learning and training with integrity is not simply reserved for Higher Education students but is vitally important for VET graduates so they can practice safely in their designated professions. The remainder of this chapter describes the two-and-a-half-year academic integrity journey at Technical And Further Education New South Wales (TAFE NSW) and the specific initiatives developed and implemented before and after the emergence of generative artificial intelligence (genAI). The narrative highlights the importance of collaboration across multiple departments to achieve a cohesive response. This chapter is based on my experience as the Academic Integrity Lead from 2022–2024 at TAFE NSW.

Academic Integrity in Australian Tertiary Education

TAFE NSW, Australia's largest education and training provider, offers over 1200 courses or qualifications with levels ranging from Australian Qualifications Framework Levels 1-8 (Department of Education, 2013) and sees more than 400,000 enrolments annually. Tertiary education in Australia, which encompasses vocational and Higher Education, is regulated by two entities established in 2021: The Australian Skills Quality Authority (ASQA) for the VET sector and the Tertiary Education Quality Standards Authority (TEQSA) for the higher education sector. Within the Australian context, the term 'Higher Education' applies only to courses that lead to a bachelor's degree or higher (Australian Qualifications Framework Level 8).

In June 2020, TEQSA opened a dedicated Integrity Unit (TEQSA, 2022b), one of its focal points being academic integrity. TEQSA has been instrumental in promoting academic integrity by providing a wide range of resources, enacting legislation against the contract cheating industry through the Prohibiting Academic Cheating Services Bill 2019 (TEQSA, 2022c) and fostering collaboration across the sector in response to generative artificial intelligence (genAI), including guidelines for rethinking assessments (TEQSA, 2024b, 2024c). Academic integrity is also a priority in high schools in the Australian state of NSW, where students must complete the "HSC: All My Own Work" program, which promotes academic integrity before starting Year 11 (NSW Government, n.d.) and before entering the High School Certificate (HSC). The HSC score forms the basis for students' entry into tertiary education. VET providers who have aligned their academic integrity expectations with those of TEQSA view academic integrity initiatives through the lens of quality education, however, there are differences in expectations across providers.

ASQA (2024) uses 'authenticity' as one of the four requirements in all assessments, which essentially calls for the same principles as academic integrity. Authenticity requires that "the assessor is assured that the evidence presented for assessment is the student's own work" (ASQA, 2024). Unlike Higher Education teachers, all VET teachers in Australia are required to complete the Certificate IV in Training and Assessment and update their certification when a new version is released. However, there are no set requirements for VET providers to mandate the completion of academic integrity modules or other educational measures. Therefore, the academic integrity discussion has followed a different trajectory among VET providers. In August 2023, ASQA mentioned 'academic integrity' for the first time in its newsletter and created a new webpage, indicating its significance and alerting readers to the perils of contract cheating services and the challenges posed by genAI (ASQA, 2023).

Elevating Academic Integrity at TAFE NSW

Incorporating academic integrity standards and requirements of TEQSA into all VET courses goes beyond meeting minimum compliance standards. TEQSA provides a benchmarking checklist for organisations to ensure they meet academic integrity standards (TEQSA, 2022a), which TAFE NSW has used. Doing so enhances the quality of educational programs in VET courses and addresses the increased societal concern over academic integrity in public education following multiple educational scandals in Australia (TEQSA, 2017). The importance of this connection should not be overlooked, as changing societal expectations is a critical marker that Higher Education must also change (Griffioen, 2023).

Although a role or unit dedicated to academic integrity has become common practice among Higher Education providers in Australia, this is non-existent for VET providers. TAFE NSW was the first VET provider in the country to create an *Academic Integrity Lead*, showing the commitment to upholding the highest standards of integrity and providing leadership across a range of responsibilities within the institution (Glendinning, 2022). The Chief Product & Quality Officer, who contributed to academic integrity research and the TEQSA academic integrity toolkit, was instrumental in creating this role. This reflects the importance of senior managers, directors, and C-suite staff in believing in the critical nature of academic integrity within the institution. The academic integrity strategies employed at TAFE NSW take a holistic approach that emphasises academic integrity is 'everyone's business' and that it is important to #makeitsomeonesjob, a hashtag popularised by Cath Ellis and Kane Murdoch, two leading academic integrity researchers and advocates.

Boundary-spanning approaches fueled the academic integrity strategies applied at TAFE NSW. Boundary spanning is a view on collaboration across institutional boundaries that can meet the challenges in today's quickly shifting Higher Education landscape (see Prysor & Henley, 2018). As such, consciously investing in boundary-spanning activities facilitates an integrated and systems-aware approach to change and avoids top-down methods that can feel forced or siloed systems that do not incorporate all stakeholders' skills (Cross et al., 2013). The remainder of this chapter describes the academic integrity pathway at TAFE NSW before and after the release of accessible genAI technology.

Stage One: Prior to the Release of Accessible GenAI

Evidence-based change practices can be used to respond to observed problems, but first, change-makers should listen and cultivate respect before presenting evidence or decisions to activate stakeholders' intrinsic motivation (Herman & Loewenstein, 2017). Therefore, it was essential to first understand the complex, large-scale organisation and structure of TAFE NSW by meeting with various stakeholder groups. Following this vital first step, presentations about the importance of academic integrity were delivered to initiate further conversation and gather feedback, experiences, and views from staff in various roles and ensure they felt part of the change process. A working group was formed where staff representing Teachers, Head Teachers, TAFE Services Managers, Regional Directors, Librarians, and Counsellors could express their views and provide input, ensuring that the community was included in the proposed changes. All staff members were invited to join a community of practice via a Microsoft Teams site where important information on academic integrity was shared, and questions were welcomed. Over 600 staff in various roles, including senior leaders, joined this network.

Following this step, evidence-based academic integrity strategies were initiated to apply best practices from researchers and practitioners in Australia and overseas. Particularly influential was a TEQSA (2020) workshop where Tracey Bretag encouraged faculty and administrators to demonstrate that academic integrity matters, not just in writing and citing, but in actions and every aspect of the work. Bretag called for clear expectations for all stakeholders with mutual participation by all, including students, and that policy must be backed up by practice, with procedures to promote good decision-making and identify and respond appropriately to issues. In order to apply these ideas and think systematically to identify interrelated elements, we relied on approaches that transcend traditional boundaries and connect stakeholders with innovative activities at multiple levels (Mirci & Hensley, 2010).

Framework for Academic Integrity

Under the guidance of the Academic Integrity Lead, TAFE NSW developed and implemented six strategies as part of the holistic institutional framework for advancing academic integrity (see Fig. 4.1).

Effective Education of Staff and Students Three primary factors contribute to breaches of academic integrity (specifically contract cheating): (a) dissatisfaction with the teaching and learning environments, (b) perception of many opportunities to cheat, and (c) speaking a language other than English at home in a primarily

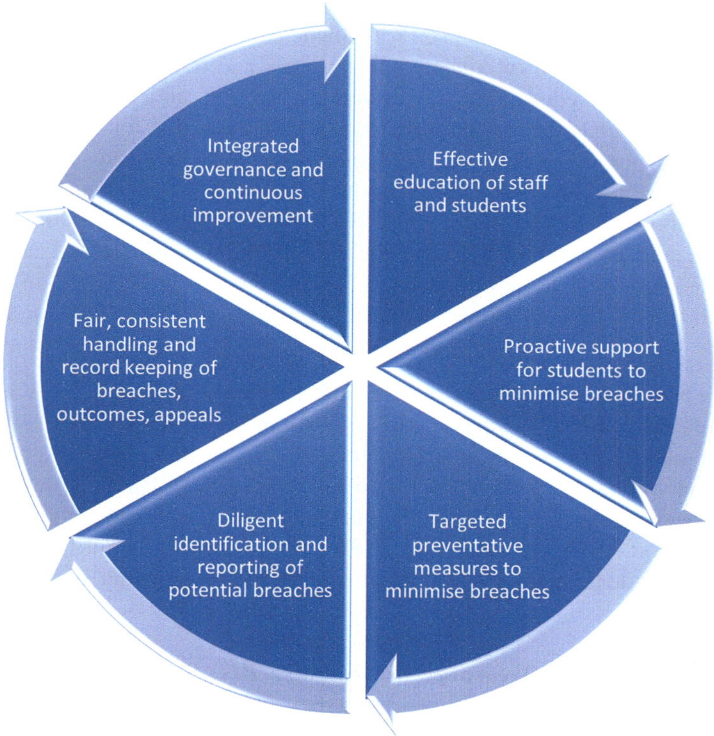

Fig. 4.1 TAFE NSW Framework for Academic Integrity

English-speaking country (Bretag et al., 2018). Students may also resort to cheating when they lack commitment and intrinsic motivation to study, have weak general academic skills (e.g., time management), perceive that staff do not value academic integrity, deal with personal or family challenges, and/or possess certain psychological traits (e.g., anxiety, low self-control, and impulsivity) (TEQSA, 2020, 53:34–1:00:00). With these research findings in mind, identifying the root causes of academic integrity breaches was an important element in developing and implementing a holistic and positive approach to academic integrity at TAFE NSW. When meeting with students to discuss breaches, underlying causes must be examined. This approach allows us to address structural issues, identify skills that students need to acquire, and refer students to appropriate support services—all to prevent future breaches.

According to Sefcik et al. (2019), education on academic integrity ensures that students and staff become competent in academic integrity. Educated staff are crucial to strengthening a culture of academic integrity because their understanding and actions shape how they guide students and how students view academic integrity. Although even a single three-hour workshop can enhance understanding of academic integrity (Curtis et al., 2021), a holistic educational approach must have a comprehensive collection of educational materials and training. Therefore, TAFE NSW developed an educational framework to support understanding academic integrity and the underlying causes of breaches. Without this foundation, smaller-scale strategies are less impactful. Academic integrity expectations and other content, including links to policy and procedures, were incorporated into the annual mandatory training module for all TAFE NSW staff, websites, orientations, course inductions, and assessments so that the message is consistent and accessible. Regular discussion of the implications of academic integrity breaches, in terms of policy and real-world consequences, further reinforces its importance for staff and students.

Proactive Support for Students to Minimise Breaches At TAFE NSW, teachers are encouraged to identify gaps in skills, closely monitor student progression, and encourage peer-to-peer and teacher-student connections in the classroom. We can reduce academic integrity breaches by promoting engaging teaching methods and providing support for assessments. This includes raising awareness about the various free support services available at TAFE NSW, asking the right questions, and listening to students with the intention of supporting learning and preventing breaches.

Targeted Preventative Measures to Minimise Breaches The design and re-design of assessments is necessary to ensure they are robust against academic integrity breaches. This involves assessments that are student-centric and of an appropriate level of difficulty. At TAFE NSW, Turnitin Feedback Studio is used as a preventative tool for students to improve their work before submitting final versions of their work. Turnitin is used in selected courses at Australian Qualifications Framework Level Four and above that were more at risk for written breaches of academic integrity or high risk to the community if graduates are not competent. Staff were trained to understand how to use Turnitin as an educative tool that empowers students. Technological solutions are deemed even more essential in fully online courses where the "know thy student" approach is less practical and relies on student participation in video conferencing sessions. Assignment Watch (n.d.), a tool that can be set up to scan the internet for assessments uploaded to commercial file-sharing sites, is used in some courses such as Diploma of Nursing. Turnitin is used in selected courses at Australian Qualifications Framework Level Four and above that were more at risk for written breaches of academic integrity or high risk to the community if graduates are not competent. Staff were trained to understand how to use Turnitin as an educative tool that empowers students. Technological solutions are deemed even more essential in fully online courses where the 'know thy student' approach is less practical and relies on student participation in videoconferencing sessions.

Diligent Identification and Reporting of Potential Breaches The institution must ensure that teachers, assessors, and invigilators are trained to identify potential breaches diligently. This includes taking adequate time to mark assignments, recognising evidence of work that may not be the student's own, and appropriately using software like Turnitin Feedback Studio and Authorship Investigator.

Fair, Consistent Handling and Record Keeping of Breaches, Outcomes, and Appeals Given the number of staff spread across 147 campuses in the state, it is vital for those who hold formal responsibilities named in the academic integrity procedure to be trained to follow it consistently. The updated policy and procedure will include training recordings, written guides supporting specific roles, and weekly Q&A sessions.

Integrated Governance and Continuous Improvement TAFE NSW's academic integrity strategy promotes accountability by updating internal governing bodies. Upon request, end-of-semester academic integrity breach data from courses is sent to the Educational Quality Committee consisting of chiefs, directors, and representatives from senior leadership. Updates on projects and necessary approvals are also shared, providing ongoing guidance and continuous improvement.

Academic Integrity Policy and Procedure

The Exemplary Academic Integrity Policy Project offered advice on developing policy and procedures that recognise the key elements of access, approach, responsibility, detail, and support (TEQSA 2020, 1:01-1:10; University of South Australia, n.d.). Using these guidelines to create a robust academic integrity policy and procedure formed the foundation of good academic integrity practice at TAFE NSW. To start, new terminology aligned with the positive and nurturing approach replaced

Table 4.1 Updated academic integrity policy and procedure language at TAFE NSW

Previous Terminology	Replacement Terminology
Student Conduct & Discipline Procedure	Academic Integrity Procedure
Academic misconduct	Academic integrity breach
Allegation	Concern/Potential breach
Penalty	Outcome
Hearing	Meeting/Conversation

older terminology (see Table 4.1). However, the resulting consequences remain unchanged and reflect the serious nature of academic integrity breaches.

The prevention of breaches is positioned at the beginning of the procedure to highlight its significance, offering links to a range of support staff and academic integrity modules. Additionally, there are details on the immediate and longer-term dangers of dealing with contract-cheating service providers and a link to report them to TEQSA. Stakeholders will also find detailed information on the types and outcomes of breaches and the appeal process. Finally, the procedure includes information on record keeping and reporting of breach data by courses at the end of the semester, as well as information on breach analysis and governance reports (as previously described).

The procedure stipulates that everyone involved, from teachers to those handling breaches, such as Head Teachers, Course Coordinators, and TAFE Services Managers, should consistently communicate with students in a respectful manner. Head Teachers should ask questions to learn more about underlying causes when discussing suspected breaches and enable the provision of targeted support to prevent future breaches. These discussions should leave the student feeling respected, cared for, and confident in the available support to develop new skills. Moreover, the choice of words in letters or emails sent to students is crucial; therefore, templates are provided to ensure consistency and promote a positive, empowering experience. This approach, which seeks to restore and educate rather than merely punish, embodies the principles of restorative justice.

Additional Supports

The Academic Integrity Lead at TAFE NSW also supports other initiatives, such as blocking contract cheating websites. TEQSA requests major Internet Service Providers to block illegal cheating websites, making them publicly unavailable. Additionally, TEQSA updates its database of suspected contract cheating services providers (currently ~ 3000; see TEQSA, 2024d) regularly so that institutions, including TAFE NSW, can block access to these sites from their networks.

The Australasian Academic Integrity Network (AAIN, n.d.) is invaluable for discussing academic integrity, benchmarking, and best practices with other practitioners, educators, and staff members who are passionate about the topic. Although participants primarily consist of Higher Education staff, the established and effective methods used in Higher Education have benefited the adaptation and integration of academic integrity into the VET context. Attendance at the yearly forums is promoted among TAFE NSW staff to plan events, attend, present, and join the network, facilitating a broader understanding and implementation of academic integrity across our institution. Attendance also builds camaraderie among professionals dedicated to promoting academic integrity.

Stage Two: Riding the GenAI Wave

The release of ChatGPT in November 2022 sparked a significant concern about academic integrity. An influx of questions about genAI was anticipated from TAFE NSW staff returning to work after the summer break, especially considering its ban in state K-12 schools. As a government organisation offering high school subjects, it was crucial for staff to understand TAFE NSW's stance and guidelines. A metaphor was used to liken GenAI technology to a venomous snake that can bite and have harmful impacts but could also provide ‘antivenom’ if administration to the bite was timely and strategic. Just as antivenom uses snake venom, genAI, used wisely and ethically, can be a cure. Together, this situation prompted urgent institutional action, and we learned that we could ride the wave of concern over genAI to promote academic integrity in new ways. The introduction of genAI sparked open discussions about the principles and expectations of academic integrity, along with potential responses, including assessment re-design, revision of policy and procedure, and the development of new and the revision of existing initiatives.

Staff Professional Development

The Academic Integrity Lead addressed the challenges and opportunities presented by genAI by demonstrating how the technology could be leveraged to create an engaging teaching and learning environment (Koren & Anders, 2024). The technology can also provide tailored assistance to meet the unique learning needs of English as an Additional Language (EAL) students, thereby addressing several known root causes of academic integrity breaches (Bretag et al., 2018). Therefore, two large professional development sessions were conducted in week 1 of Semester 1 2023—one on using Gen AI to minimise breaches and support educators and students and the other on ethical considerations and guardrails.

Collaboration between individuals and teams with diverse roles within the organisation was also essential. Reviewing the responsibilities and teaching background of the Skills Team Education Excellence Support (STEES), it became evident that they were critical for supporting teachers using genAI tools. Recognising the large number of teachers, potential reluctance to try new tools, and the fact that specific genAI tools are better suited to certain industries and fields of study, the Academic Integrity Lead indicated the need for training tailored to each Skills Team. The Managers of Education Quality organised a team of STEES aligned with each Skills Team to conduct genAI and academic integrity training for teachers. The Academic Integrity Lead was available to answer questions posed by teachers during the training. During each presentation, academic integrity expectations were shared, with links to the academic integrity intranet, student-facing webpages, academic integrity module, and platform for facilitating regular engagement between community members.

Another active team is the Learning Technology Services Team, which is responsible for integrating new technologies into teaching and training staff. The Education Technology Lead organised a TAFE-wide roundtable discussion on genAI, which was open to all, facilitating an excellent opportunity for networking and collaboration across teams. Since October 2023, the Learning Technology Services Team have conducted genAI training sessions focused on how to use specific tools effectively.

Policy and Procedure

AAIN member institutions have categorised the use of genAI in a variety of ways, such as plagiarism, inappropriate use of genAI, or contract cheating. According to TEQSA (2023), however, "artificial intelligence (AI) software or paraphrasing tools can be considered a form of contract cheating, where substantial parts of exam questions and assignments are written by text-generating software." The TAFE NSW academic integrity policy and procedure documents were updated in March 2023 to include the phrase "inappropriate use of genAI" (see also Monash University, 2023).

Teacher and Student Guides

Clear guidelines on the acceptable use of genAI for educators and students were necessary for consistent messaging. Consequently, the Academic Integrity Lead developed the *Student Guide to ChatGPT* and the *Teacher Guide to ChatGPT* in February 2023 (later converted to infographics and renamed) to promote the appropriate use of these tools in teaching and learning. The 2023 and *2024 TAFE NSW Student Guides* (TAFE NSW, 2024a) and the 2023 and *2024 TAFE NSW Mandatory Training* were updated to include academic integrity content and the link to the *GenAI and Academic Integrity Library Guide* (TAFE NSW, 2024b). Examples of content in these guides are featured below.

Competency Conversations An important feature of student and teacher guides is the discussion on verifying that submitted work reflects students' competencies. In a competency conversation, teacher meets with a student to discuss their submitted content and confirm their understanding if there is a concern about a potential academic integrity breach. Teachers have a guide on how to lead a competency conversation, and students learn that they might be asked to discuss their submitted content and present their research and drafts.

GenAI to Provide Feedback The teacher guide indicates that teachers should not use genAI to assess student work for three primary reasons: (1) educators must observe repeated errors or misunderstandings to offer appropriate assistance or additional instruction; (2) genAI is not always accurate, and human review is crucial; and (3) being aware of student's knowledge, skills and progress and providing personalised feedback ensures that students feel appreciated and benefit from the teacher's personalised and thoughtful comments. Further research is required to determine the effectiveness of AI-generated feedback; however, teachers may explore how prompts can be crafted to generate empathetic and engaging feedback without sharing identifiable student information (Koren & Anders, 2024). If genAI is developed within the organisation and trained on institutional teaching materials in the future, feedback from genAI, along with teacher feedback, could increase efficiencies for educators.

GenAI Detection When Turnitin released its AI detection tool in April 2023, training sessions were organised for staff at TAFE NSW to learn how to interpret the results with care. We emphasised that the percentage or highlighted portion was not clear evidence of the presence or absence of a breach and that teachers should engage students in conversation about the submitted assessment. In January 2024, the genAI detection tool was disabled due to the risk of false positives and false negatives. GenAI detection may be reenabled if the report is integrated into the Authorship Investigator dashboard, where only the investigator can view it.

Library Support

The library team is an essential group that supports academic integrity, especially with research and referencing. The two library representatives on the academic integrity working group—the Manager of Library Services and the Library Network Team Lead, offered to host student genAI and academic integrity resources on the library webpage. The genAI resource (TAFE NSW, 2024b) was developed to support a clear understanding of genAI, featuring numerous external links for further exploration of its various aspects, FAQs, and additional resources, including TAFE NSW's guides, academic integrity module, policies, and procedures. One challenge was determining the appropriate way to acknowledge the use of genAI because formal style and documentation guides did not include genAI. Later, the American Psychology Association (APA) advised referencing genAI use as personal communication but later suggested referencing it as a software tool (McAdoo, 2024).

Since August 2023, the library team has also conducted regular sessions for students about the appropriate use of genAI tools and the promotion of academic integrity. The collaboration with the library team has led to presenting the TAFE NSW genAI response at an Australia-wide event for libraries. This highlights that forming partnerships across an institution can expedite the development and implementation of solutions when issues arise.

Academic Integrity Modules

Student Module In May 2023, the Academic Integrity Lead developed an academic integrity module to include in-depth information about academic integrity and a section on genAI, which takes up to two hours to complete. Collaborating with the TAFE NSW state-wide orientation team, this module was presented to Head Teachers with four slides defining academic integrity, why it is important, and the benefits for the student, TAFE NSW, and the community. Students are introduced to the module and recommended to complete it during the Course Induction in a group session led by a Head Teacher. Head Teachers were advised to present the slides, contextualised to their course, with course-specific examples, examples of past breaches, and consequences and their impact on their industry when graduates lack integrity. They were advised to complete the academic integrity module in person in group discussions to play a game to ensure students retain the content. Live peer-to-peer and peer-staff interactions are more likely to engage the students, allowing them to express their views and ask questions, which better supports their learning about academic integrity and contract cheating (Lancaster, 2022).

Currently, the completion of the module is not mandated in VET courses but only for Higher Education courses, and the e-learning platform is configured to restrict access to Higher Education assessments until module completion. This decision may, however, be revisited in the future. Moreover, Higher Education teachers who cover first-year academic skills subjects were advised to discuss the module content after its completion, contextualising academic integrity within the students' broader educational and professional lives. Highlighting the risks of academic integrity breaches, such as graduating without the intended learning outcomes, highlights the potential long-term implications of engaging in inappropriate academic behaviours (Glendinning, 2022).

Staff Module A staff academic integrity module, scheduled for release in mid-late 2025, discusses the importance of academic integrity and what it means beyond identifying breaches, focusing on how teachers, librarians, disability services, learning centre staff, product development teams, and senior leaders can foster cultures of academic integrity. Completing the Staff Academic Integrity Module will be recommended for all staff; however, it is prioritised for teachers, Head Teachers, and those involved in handling breaches and appeals. Module completion is to be integrated into the Teacher Individual Development Plan (IDP), allowing Head Teachers to track completion. The Academic Integrity Lead invited various staff to describe examples of exemplary practice in short videos to be included in the module so that participants can be inspired by their peers. The module also includes practical examples and links for further reading, offering a comprehensive yet concise resource that encourages further exploration. Furthermore, the module demonstrates how the efforts to promote academic integrity align with the organisation's strategic plan and are integral to the staff code of conduct, as advised by TEQSA (2020, 2:02:20–2:03:31).

Academic Integrity Communications

Academic Integrity Webpage TEQSA's Standard 7.2.2 (TEQSA, 2024c, d) requires that prospective students be made aware of the academic integrity policy (TEQSA, 2024a). Although ASQA does not have a similar requirement for VET, TAFE NSW demonstrates its commitment to higher quality by adhering to TEQSA's standards. TAFE NSW has prioritised the academic integrity webpage as an essential resource, as it is designed to ensure that the institution's academic integrity expectations are clearly communicated with easy access to policy, procedure, and academic integrity modules (Bretag et al., 2018). The webpage's tone portrays a welcoming and student-centric approach. Readers will have access to the academic integrity framework (detailed earlier in this chapter), details on the student's immediate and long-term personal gains when following principles of academic integrity, and links to the student academic integrity module, institutional policy, and free support services that can assist with good academic practice, all in an easily accessible location.

Staff Intranet A centralised collection of academic integrity resources was developed on a SharePoint intranet page in March 2023, providing staff with a single location to access all academic integrity materials. The intranet is designed to instil the importance of academic integrity, applying the TAFE NSW vision of being "a leading provider of lifelong learning, meeting the evolving needs of industry and learners in communities across NSW" (TAFE NSW, 2024c, p. 7) along with the vision for academic integrity as "the cornerstone of lifelong education, nurturing lasting impact on learners, community, industry, and our reputation" (TAFE NSW Intranet). As a centralised hub, staff can find all the details about the academic integrity framework explained, module links, slides on academic integrity to be used in classes, calendar links to weekly Q&A sessions, upcoming academic integrity-focused events, recordings of past professional development sessions, links to policy, procedure, and support materials, and FAQs. Staff will also find one-page guides for each of the staff roles responsible for the procedure, such as teachers, Head Teachers, Course Coordinators and TAFE Services Managers, along with templates for the stages of contact with students regarding breach concerns through to outcome and appeal letters, and end of term reporting. Staff are encouraged to ask questions through forums on the intranet, and they will find links to join the TAFE NSW academic integrity network for ongoing discussions and the weekly Q & A drop-in sessions for emerging questions. Other vital institutional resources to support students' holistic well-being, such as personal and career counselling and disability services, are also posted.

Managing Director's Newsletter Since the end of January 2023, the individual who holds the highest leadership role at TAFE NSW has communicated the institutional response to genAI through the Managing Director's newsletter, reaching a wide audience. For example, the newsletter has announced professional development sessions, the Academic Integrity and Gen AI resource page for students, the academic integrity module and the guides for students and instructors. The newsletter also directed staff and faculty to the updated policies and procedures in response to genAI and provided updates on the intranet page. By sharing these developments, the Managing Director has emphasised the importance of academic integrity, reaching beyond teaching teams to engage senior leaders. This strategy enabled more staff to receive important updates and underscored the critical role that every institution member plays in fostering academic integrity. The Managing Director's approach is exemplary, showcasing how top-down support is fundamental for promoting and sustaining an institutional culture of academic integrity.

Conclusion

This chapter illustrated the two-and-a-half-year journey to shift perspectives on academic integrity in VET, starting with a dedicated practitioner role. The Academic Integrity Lead has ensured that efforts to build a robust academic integrity strategy are coordinated and sustained with engagement from all stakeholders. Collaborative, boundary spanning approaches leveraged the expertise of colleagues throughout the organisation to enhance the collective institutional response. The benefits of facilitating strong inter-departmental relationships became evident as stakeholders worked together to support academic integrity when new challenges were presented, such as the pivotal moment when ChatGPT and other accessible genAI tools exploded since November 2022. Framing genAI as a snakebite that threatens academic integrity standards and the antivenom that strengthens academic integrity responses has been a memorable analogy in the Australian context. Through evidence-based practices and collaborative initiatives, TAFE NSW has demonstrated that VET institutions can grow a strong institutional culture for academic integrity by implementing a holistic framework with strategies that meet or exceed regulatory requirements. By documenting the experience at TAFE NSW, this chapter aimed to inspire other VET and polytechnic providers to establish or strengthen their academic integrity practices.

Acknowledgements It is fitting to acknowledge the exemplary role played by TEQSA's Integrity Unit, which developed a valuable range of resources and responses to genAI and contract cheating, as well as standards for education providers. I recognise the significant influence of the late Dr. Tracey Bretag, whose research and passionate commitment to academic integrity inspired the strategies and initiatives pursued in the role of Academic Integrity Lead. Finally, I extend my gratitude to all staff at TAFE NSW whose assistance and engagement have been fundamental in successfully implementing academic integrity projects. Their collective efforts, insightful feedback on practicality based on the organisational structure, and commitment to advancing academic integrity have been crucial. Each big and small contribution has enriched this endeavour and impacted our collective achievements significantly.

References

Assignment Watch. (n.d.). https://assignmentwatch.com/

Australasian Academic Integrity Network. (n.d.). *About us*. https://academicintegrity.edu.au/about/

Australian Skills Quality Authority [ASQA]. (2023, August 30). *Academic integrity*. Australian Skills Quality Authority; Australian Government. https://www.asqa.gov.au/how-we-regulate/risk-priorities/academic-integrity

Australian Skills Quality Authority [ASQA]. (2024, March 1). *Clauses 1.8 to 1.12 — conduct effective assessment*. Australian Skills Quality Authority; Australian Government. https://www.asqa.gov.au/rtos/users-guide-standards-rtos-2015/chapter-4-training-and-assessment/clauses-18-112-conduct-effective-assessment

Bretag, T., Harper, R., Burton, M., Ellis, C., Newton, P., Rozenberg, P., Saddiqui, S., & van Haeringen, K. (2018). Contract cheating: A survey of Australian university students. *Studies in Higher Education, 44*(11), 1–20. https://doi.org/10.1080/03075079.2018.1462788

Cross, R., Ernst, C., & Pasmore, B. (2013). A bridge too far? How boundary spanning networks drive organizational change and effectiveness. *Organizational Dynamics, 42*(2), 81–91. https://doi.org/10.1016/j.orgdyn.2013.03.001

Curtis, G. J., Slade, C., Bretag, T., & McNeill, M. (2021). Developing and evaluating nationwide expert-delivered academic integrity workshops for the higher education sector in Australia. *Higher Education Research & Development, 41*(3), 1–16. https://doi.org/10.1080/07294360.2021.1872057

Department of Education. (2013, January 16). *AQF second edition*. Australian Qualification Framework. https://www.aqf.edu.au/publication/aqf-second-edition

Glendinning, I. (2022). *Aligning academic quality and standards with academic integrity*. In S. E. Eaton, G. J. Curtis, B. M. Stoesz, J. Clare, K. Rundle, & J. Seeland (Eds.), *Contract cheating in higher education: Global perspectives on theory, practice, and policy* (pp. 199–218). Palgrave Macmillan. https://link.springer.com/book/10.1007/978-3-031-12680-2

Griffioen, D. M. E. (Ed.). (2023). *Creating the desire for change in higher education*. Bloomsbury Academic. https://library.oapen.org/bitstream/handle/20.500.12657/58875/1/9781350244771.pdf#page=58

Herman, G. L., & Loewenstein, J. (2017). Evidence-based change practices. *Journal of Engineering Education, 106*(1), 4–13. https://doi.org/10.1002/jee.20152

Koren, N., & Anders, B. (2024). Harnessing the power of generative artificial intelligence to promote academic integrity. In S. Mahmud (Ed.), *Academic integrity in the age of artificial intelligence* (pp. 241–262). IGI Global. https://doi.org/10.4018/979-8-3693-0240-8.ch013

Lancaster, T. (2022). Addressing contract cheating through staff-student partnerships. In S. E. Eaton, G. J. Curtis, B. M. Stoesz, J. Clare, K. Rundle, & J. Seeland (Eds.), *Contract cheat-*

ing in higher education: Global perspectives on theory, practice, and policy (pp. 219–232). Palgrave Macmillan. https://link.springer.com/book/10.1007/978-3-031-12680-2
McAdoo, T. (2024). *How to cite ChatGPT.* APA Style. https://apastyle.apa.org/blog/how-to-cite-chatgpt
Mirci, P. S., & Hensley, P. A. (2010). *Leading for innovative practice: Melding theories of organizational change, adult learning, and conditions of learning*. CAPEA Education Leadership and Administration, 22. https://files.eric.ed.gov/fulltext/EJ965159.pdf
New South Wales (NSW) Government. (n.d.). *HSC: All my own work program*. NSW Government; NSW Education Standards Authority. Retrieved April 29, 2024, from https://educationstandards.nsw.edu.au/wps/portal/nesa/11-12/hsc/hsc-all-my-own-work
Prysor, D., & Henley, A. (2018). Boundary spanning in higher education leadership: Identifying boundaries and practices in a British university. *Studies in Higher Education, 43*(12), 2210–2225. https://doi.org/10.1080/03075079.2017.1318364
Sefcik, L., Striepe, M., & Yorke, J. (2019). Mapping the landscape of academic integrity education programs: What approaches are effective? *Assessment & Evaluation in Higher Education, 45*(1), 30–43. https://doi.org/10.1080/02602938.2019.1604942
Technical and Further Education New South Wales [TAFE NSW]. (2024a). *Student Guide.* Student Services. https://www.tafensw.edu.au/documents/60140/294441/TAFE-NSW-Student-Guide.pdf
Technical and Further Education New South Wales [TAFE NSW]. (2024b). *Researching and referencing: Generative AI.* Library Services.. https://tafensw.libguides.com/research/artificial
Technical and Further Education New South Wales [TAFE NSW]. (2024c). *TAFE NSW Strategic Plan 2022-25.* https://www.tafensw.edu.au/documents/60140/76288/TAFE-NSW-Strategic-Plan-22-25.pdf
Tertiary Education Quality and Standards Agency (TEQSA). (2020, February 13). *Academic integrity in Australian higher education: A national priority*. YouTube. https://youtu.be/E3-IaTlZdR8?si=60ZhV3epZqFKGxhI
Tertiary Education Quality and Standards Agency (TEQSA). (2022a). Self-evaluation and benchmarking checklist for academic integrity key principles. In *Policy and benchmarking*. https://www.teqsa.gov.au/sites/default/files/2022-10/self-evaluation-benchmarking-checklist-web.pdf
Tertiary Education Quality and Standards Agency (TEQSA). (2022b, October 13). *Higher education integrity unit.* Tertiary Education Quality and Standards Agency. https://www.teqsa.gov.au/about-us/teqsa-overview/higher-education-integrity-unit
Tertiary Education Quality and Standards Agency (TEQSA). (2022c, October 13). *Legislation.* Tertiary Education Quality and Standards Agency; Tertiary Education Quality and Standards Agency. https://www.teqsa.gov.au/preventing-contract-cheating/legislation
Tertiary Education Quality and Standards Agency (TEQSA). (2023, January 17). *Sector update: Maintaining up to date academic integrity policies and procedures* | Tertiary Education Quality and Standards Agency. Teqsa.gov.au. https://www.teqsa.gov.au/sector-update-maintaining-academic-integrity-policies-and-procedures
Tertiary Education Quality and Standards Agency (TEQSA). (2024a, February 24). *Guidance note: Academic and research integrity | tertiary education quality and standards agency.* Teqsa.gov.au. https://www.teqsa.gov.au/guides-resources/resources/guidance-notes/guidance-note-academic-and-research-integrity
Tertiary Education Quality and Standards Agency (TEQSA). (2024b, February 26). *Protecting academic integrity. Tertiary Education Quality and Standards Agency*; Tertiary

Education Quality and Standards Agency. https://www.teqsa.gov.au/guides-resources/protecting-academic-integrity

Tertiary Education Quality and Standards Agency (TEQSA). (2024c, March 26). Artificial intelligence. Tertiary Education Quality and Standards Agency; Tertiary Education Quality and Standards Agency. https://www.teqsa.gov.au/guides-resources/higher-education-good-practice-hub/artificial-intelligence

Tertiary Education Quality and Standards Agency (TEQSA). (2024d, May 28). Blocked illegal cheating websites. https://www.teqsa.gov.au/blocked-illegal-cheating-websites#list-of%20blocked%20illegal%20cheating%20websites

Tertiary Education Quality and Standards Agency [TEQSA]. (2017, October). *Good practice note: Addressing contract cheating to safeguard academic integrity.* https://www.teqsa.gov.au/sites/default/files/good-practice-note-addressing-contract-cheating.pdf

University of South Australia. (n.d.). *Exemplary academic integrity project.* https://lo.unisa.edu.au/course/view.php?id=6751&sectionid=112510

Nadia Koren is the Statewide Academic Integrity Lead for TAFE NSW, the largest post-secondary education and training provider in the Southern Hemisphere, primarily focused on Vocational Education and Training (VET) while also offering Higher Education and school-level programmes. She holds a Bachelor of Science (Computer Science) and a Master of Applied Science (Information Technology). Over her 17-year career in tertiary education, she has worked for 5 years in Vocational Education and 12 years in Higher Education, both as an educator and in various leadership roles, with 5 of those years specifically dedicated to academic integrity work. She is passionate about student engagement, assessment design, and integrating generative AI into educational practices to improve outcomes for students. Nadia promotes a holistic approach to education, focusing on the development of both cognitive and emotional skills for lifelong learning, and advocates for a restorative justice approach to academic integrity. She was a finalist for the 2024 Tracey Bretag Global Academic Integrity Award, recognised for her scholarly approach in responding to generative AI, developing resources, and adopting a whole-of-organisation strategy to promote and drive a positive culture shift in academic integrity using generative AI.

Chapter 5
Faculty Education on Academic Integrity: A Micro-Credential Approach

Cory Scurr

Abstract The need for academic integrity education is clear, and this has been well documented; however, much of the literature focuses on the importance of educating students. Faculty hold a significant role in integrating academic integrity in course delivery, but they may not be attuned to academic integrity expectations and the benefits an academic integrity framework provides from a teaching and learning perspective. Faculty in vocational and polytechnic institutions typically come from industry and may have no previous teaching experience. These subject matter experts can be allies and champions for promoting and maintaining academic integrity by connecting the values of academic integrity to the values of their profession. Educating and empowering vocational and polytechnic faculty ensures a wide reach for integrity values and should be top of mind.

One optimal way to build rich academic integrity awareness at an institutional level is to educate faculty via a micro-credential. This chapter explores one Canadian polytechnic's academic integrity micro-credential framework and examines what precipitated its initiation and what drives its focus. A "sneak peek" into each of the five courses (*An Overview of Academic Integrity: Building a Culture of Integrity; The Awkward Conversation: Barriers to Reporting Academic Misconduct & Talking to Students; An Inherent Conflict? Academic Integrity and Technology; Identifying Plagiarism & Maintaining Academic Honesty;* and *Contract Cheating in Higher Education*) and their general themes are explored.

Lastly comes a gentle nudge to vocational and polytechnic institutions to consider developing and delivering faculty education on academic integrity, leveraging a model most suitable for them. While this model for faculty education in academic integrity is delivered within a Canadian college/polytechnic context, the benefits, observations, and recommendations are relevant in all contexts since academic integrity education has no borders.

C. Scurr (✉)
Conestoga College, Kitchener, ON, Canada
e-mail: cscurr@conestogac.on.ca

L. Vogt, B. M. Stoesz (eds.), *Academic Integrity in Vocational and Polytechnic Education*, Ethics and Integrity in Educational Contexts 8,
https://doi.org/10.1007/978-3-031-90091-4_5

Keywords Academic integrity · Faculty education · College instructor · Micro-credential · Integrity education · Post-secondary · Higher education · College · Polytechnic · Vocational · Canada

The importance of academic integrity education is clear, and this is well supported in the literature by extensive evidence (Bertram Gallant, 2008, 2017; Bertram Gallant & Drinan, 2008; McCabe et al., 2004). Much of the literature focuses on the importance of educating students; however, more must be done to address the gap in preparing and supporting faculty. Many colleges and polytechnics in Canada offer training to prepare new faculty to transition from roles as industry professionals to instructors, and completing these courses may be a requirement of employment. Conestoga College in Ontario, Canada, leverages this opportunity by offering five short courses on academic integrity that can be stacked to receive the *Academic Integrity in Post-Secondary Teaching* micro-credential (https://continuing-education.conestogac.on.ca/micro-credentials/M1056), which can also be applied toward the requirements of the Post Secondary Teaching Certificate (https://tlconestoga.ca/faq-for-courses-and-credentials-in-post-secondary-teaching-and-learning/). Educating and empowering faculty to consider the implications of their decisions regarding issues related to academic integrity should be a priority to ensure an institutional culture of academic integrity.

My role is the Associate Director of the Academic Integrity Office, which provides guidance, advice, and recommendations to faculty, administrators, and students on academic integrity-related matters. My team and I provide support and resources to assist faculty in understanding the institution's academic integrity policy, procedure, and best practices since decisions are largely in their hands. This office falls within the broader portfolio of the College's Library and Learning Services. However, the micro-credentials were initiated by the Department of Teaching and Learning, which resulted in a collaboration between the two departments.

Creating a micro-credential focused on academic integrity has proved to be an optimal method to build rich academic integrity awareness at an institutional level. Faculty learn about academic integrity and misconduct through fulsome discussions, self-directed activities, and practical assessment tasks. Upon completion, participants have gained a toolkit for navigating what is often a daunting element of teaching. In this chapter, I describe the framework for the five-part micro-credential. The course structure and learning outcomes are detailed, along with a sneak peek at the themes discussed in each course. Our model can serve as an example for vocational and polytechnic institutions looking to develop and deliver faculty education using an approach most suitable for their context.

Why the Micro-Credential?

In a 2022 blog post for the International Center for Academic Integrity (ICAI), Tricia Bertram Gallant drew a parallel between academic integrity and gardening, suggesting that academic integrity, like plants, needs essential elements, like good soil, water, and fertilizer. Gallant refers to Moss Kanter's "When a thousand flowers bloom," which argues that "in order for [an] innovation to root within an organization, the environment must be intentionally structured and prepared to welcome the innovation." Gallant suggests academic integrity needs a constant attendant, someone to help nurture it and provide the right environment in which it can grow. The micro-credential courses, then, are intended to nurture an institutional environment in which strong academic integrity practices will grow and flourish, and the work is guided by the academic integrity office, situating itself firmly as the gardener.

The driving force behind creating the micro-credential was to fill a gap since workshops were not providing the necessary, and indeed sought-after, depth into academic integrity-related topics. Academic integrity training at the school and program levels tends to take a high-level approach aimed at new faculty and covers topics such as policy, procedure, and overall expectations, yet many seek a deeper, more nuanced understanding of topics captured under the broader academic integrity umbrella. While targeted workshops are provided regularly throughout the semester to schools, programs, and faculty groups, the micro-credential provides something workshops and general training cannot offer: The credibility of certification.

The Micro-Credential in Context

According to Conestoga's Teaching and Learning Website, the micro-credential is "a certification confirming that the recipient has completed one or more courses that are focused on skill(s) or competency(ies) attainment" (Conestoga College, n.d.). Explained more fully:

> Micro-credentials are short, concentrated groups of courses (typically composed of five) that are flexible, innovative, timely, and based on industry needs. They are designed to be high-quality and meaningful credentials. [...] Some micro-credentials may be stackable and can be combined to form a part of a larger credential (para. 2).

Overall, micro-credentials offer a valuable opportunity for individuals to gain specialized skills and competencies that align closely with professional development.

The *Academic Integrity in Post-Secondary Teaching* micro-credential is comprised of five courses. Each course is equivalent to one course credit, so full completion of all five courses awards five course credits and a completed micro-credential. The micro-credential is offered as an elective in the Post-Secondary Teaching Certificate, which requires 30 course credits for completion. Since it is an optional component, participants are not required to register for all five courses, allowing faculty to choose which topics they would like to explore, and individual courses still contribute to the larger Post-Secondary Teaching Certificate. Conestoga employees may take the courses for free, and those outside the institution may take the courses for a fee.

The Post-Secondary Teaching Certificate has six program learning outcomes (PLOs), and four are incorporated across the five micro-credential courses. The six PLOs are as follows:

1. Work effectively within a teaching role in post-secondary settings that align with the expectations of outcomes-based and/or competency-based education.
2. Assess student learning in ways that are authentic, valid, reliable, student-centered, and aligned with Universal Design of Learning (UDL) principles.
3. Provide evidence-based, student-centered teaching that is sensitive to student experiences, values, beliefs, and needs.
4. Provide an inclusive and culturally relevant learning environment with learning opportunities that support the success of a diverse student population.
5. Utilize contextually appropriate and accessible educational resources and technologies in order to support student learning.
6. Reflect on aspects of teaching to support the ongoing development of one's identity as a post-secondary educator.

The Academic Integrity in Post-Secondary Teaching micro-credential addresses PLOs 1, 3, 4, and 6, and although PLOs 2 and 5 are not explicitly identified, student-centred UDL principles and reflection on teaching are embedded in the courses.

Each individual course has a course description, unit learning outcomes (ULOs), and course learning outcomes (CLOs). While I am responsible for developing and delivering the courses as the Associate Director of Academic Integrity, the Teaching and Learning Department and the Curriculum Planning and Operations Department review and approve the course elements. Participants are required to complete an assessment of learning, demonstrating the course learning outcomes, which is graded as a pass or fail. Receiving an assessed credential helps validate the knowledge gained by participants and provides evidence of learning, ultimately offering more credibility than participation in workshops.

Courses are taught virtually, each totalling six hours over two weeks. Every course has an online synchronous component and an asynchronous component. The online synchronous component is divided into two one-and-a-half-hour sessions on a specific day, with the days and times varying semesterly. The remaining three hours of asynchronous learning depend on the particular course and consist of readings (e.g., scholarly works, blogs), videos (e.g., TED Talks, webinars), and resources or website exploration. The asynchronous portion may be completed at the participant's convenience but is expected to be done before the assessment is submitted. The components are updated or swapped out at my discretion as new and updated material becomes available. Those interested in creating a micro-credential, or something similar, can have a lot of fun selecting which resources to include, as the selections impact the narrative and learning that occurs.

Assessment Made Meaningful

For all the courses, with the exception of *The Awkward Conversation: Barriers to Reporting Academic Misconduct & Talking to Students,* the assessments have been intentionally designed so that participants have something tangible to use in the classroom. Applying UDL principles, participants are allowed to demonstrate their learning in a variety of ways, formats, or presentation styles. For instance, they may submit their work in written format (plain text), PowerPoint, video, H5P activity (embedded in the LMS), cartoon/comic, infographic, or something entirely different. The intention is for participants to use the assessments as an opportunity to develop practical academic integrity-related resources to use with their students. The results have exceeded expectations, as some submissions were truly remarkable, innovative, and inspiring. Since the completed assessments can be used in their teaching, participants have expressed that the tasks are worth their while. I invite participants to share their assignment submissions within our discussion forum to provide a venue for demonstrating their work and learning from the work of others. I include the following statement in the assessment description and the discussion forum: "By including your submission in the discussion forum, you're giving permission to others to use and/or adjust it for their use. It is my hope that we accumulate a large bank of resources so we can all use them at appropriate times, including myself!" Assigning an assessment that results in a useful takeaway is one of the strengths of this micro-credential, and I highly encourage others interested in developing similar courses to integrate this approach.

Sneak Peak: The Five Courses at a Glance

This section will provide an overview of all five courses:

- An Overview of Academic Integrity: Building a Culture of Integrity
- The Awkward Conversation: Barriers to Reporting Academic Misconduct & Talking to Students
- Identifying Plagiarism & Maintaining Academic Honesty
- An Inherent Conflict? Academic Integrity and Technology
- Contract Cheating in Higher Education

While this is not an exhaustive examination, it will provide a clear picture of each course and its motivations. The benefits of each course will be explained, and how each contributes to an institutional culture of integrity will be highlighted.

An Overview of Academic Integrity: Building a Culture of Integrity

While the courses are not designed to be taken in a specific order, nor do any require a prerequisite, this course is the first to be offered in the semester and is occasionally offered twice a term. The sequence of the other courses is offered randomly while striving to be delivered once per term. This course introduces the simple fact that promoting integrity in the classroom is not a straightforward or one-size-fits-all approach. Instead, it involves various complex and interconnected aspects that require different approaches and considerations to be effective. The course aligns with the following Post-Secondary Teaching Certificate program learning outcomes:

- Provide evidence-based, student-centered teaching that is sensitive to student experiences, values, beliefs, and needs.
- Reflect on aspects of teaching to support the ongoing development of one's identity as a post-secondary educator.

Course Description

This course examines current philosophies and frameworks of academic integrity. Participants explore factors that may influence a student to cheat and discuss the importance of building a culture of integrity in the classroom and the institution. Finally, they review best practices and how to leverage institutional resources to maintain academic honesty and excellence.

Course and Unit Learning Outcomes

1. Discuss various reasons why a student might cheat to develop strategies for mitigating misconduct:
   - Explain course-specific reasons a student may cheat.
   - Explain factors unrelated to the course that may cause a student to cheat.
2. Develop methods of building and promoting a culture of integrity within the classroom to encourage authentic student learning:
   - Consider how Self-Determination Theory may be leveraged in the classroom to promote integrity.
   - Assess student services that help maintain academic honesty.
   - Recognize faculty supports to help identify academic misconduct.

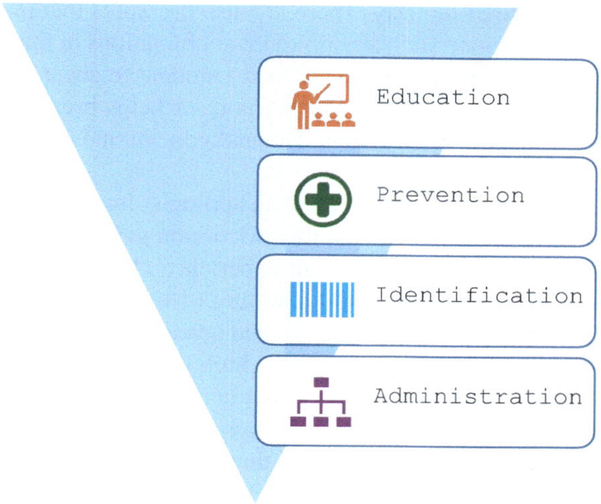

Fig. 5.1 Inverted Academic Integrity Pyramid. *Note.* Adapted from "Academic integrity for faculty library guide," by Assiniboine College (n.d.), https://assiniboine.libguides.com/ai-faculty#s-lg-box-16368504

Synchronous Sessions

The stage is set by examining the inverted academic integrity pyramid in Fig. 5.1, adopted from the Assiniboine College Academic Integrity page on their Library website (https://assiniboine.libguides.com/ai-faculty).

Introducing the inverted academic integrity pyramid highlights the dominance and importance of education and prevention to our institution's philosophical framework of academic integrity. By comparison, we focus on identification and administration less. I suggest to participants that educating faculty is as important as educating students since they are on the front lines and can positively influence student behaviour. This is also a good opportunity to ensure participants know the courses are not aimed at teaching how to *spot* cheating, although some aspects of the courses do address this. In other words, we strive to put most of our energy into academic integrity education and the least amount of energy into the administration component.

During synchronous sessions, factors influencing cheating behaviour are discussed, and faculty are asked to join small breakout rooms to reflect on these reasons. Participants have free range to generate as many reasons as possible or home in on one specific reason and unpack it; we then return to the larger group and discuss. Egan (2018), notes reasons for cheating like poor time management, peer pressure, lack of understanding, disinterest, pressure to succeed, and perception that it does not matter to the faculty. Taking a macro-environmental perspective, we also consider that students may view their education from a commercialized or transactional perspective. Kezar and Bernstein-Sierra (2016) suggest student behaviour may mimic that of the larger capitalist society (cutthroat, exploitative, and success measured quantifiably), thus signalling to students that compromising integrity is not only acceptable but necessary, which ultimately impacts the prevalence of academic dishonesty. There are also pressures that may lead to cheating outside of the classroom, such as employment obligations or family responsibilities. Exploring the myriad factors that may lead a student to cheat helps set the tone that academic integrity is not siloed. Furthermore, it helps break down barriers between faculty and students by situating students' educational experiences within a broader, more holistic framework.

The intent is not to create an apologist framework or an atmosphere wherein we forgive cheating. Rather, this discussion generates a springboard to engender common ground because we all experience challenges. It encourages faculty to begin strategizing how to build a culture of trust in their classroom that is conducive to positive learning experiences and academic integrity. We discuss the importance of interpersonal skills and, from a high level, self-determination theory (Gagne, 2014; Wade, 2017), which argues that humans need three psychological elements to feel motivated: autonomy, relatedness, and competence. I then guide a brainstorming activity on how to encourage these within the classroom.

Course Assessment

The assessment for this course is titled *Three Course Announcements/Posts,* and participants must submit three separate course announcements or posts that they could use to communicate with their students. For each announcement or post, participants must explain when they would post it (e.g., the first week of term, two days before a quiz or test, or two weeks before the final capstone project). Then, they must explain *how* the post/announcement ties to academic integrity and benefits students. Finally, they must explain *why* they are posting it.

The Awkward Conversation: Barriers to Reporting Academic Misconduct and Talking to Students

The overarching goal of this course is to provide participants with a safe space for open dialogue and discussion, guided by a facilitator with rich experience navigating difficult conversations with students and consulting with faculty. Moreover, it provides an opportunity for participants to discuss some of the more uncomfortable elements of navigating misconduct and provides constructive solutions for moving forward. What sets this course apart from the others is the amount of talking done by the participants. This course used to be offered last in the lineup; however, I quickly learned that participants wanted, and frankly needed, it offered sooner since it addresses concerns and situations faculty encounter throughout the semester, not just at the end of term. The course aligns with the following Post-Secondary Teaching Certificate program learning outcomes:

- Provide evidence-based, student-centered teaching that is sensitive to student experiences, values, beliefs, and needs.
- Reflect on aspects of teaching to support the ongoing development of one's identity as a post-secondary educator.

Course Description

This course examines academic integrity from the faculty's perspective. Participants explore potential barriers to pursuing cases of academic misconduct and the feelings associated with navigating integrity violations. They consider options when academic misconduct is suspected and strategize on how faculty can be supported as they journey through the uneven and often daunting terrain of academic misconduct.

Course and Unit Learning Outcomes

1. Evaluate potential barriers to reporting academic integrity violations to learn how to overcome them:
    - Evaluate emotional/psychological barriers to reporting academic integrity violations.
    - Appraise practical/institutional barriers to reporting academic integrity violations.
2. Determine strategies for talking with students suspected of cheating or have cheated to effectively navigate the process:
    - Interpret various types of student reactions when informed of potential academic misconduct.

Synchronous Sessions

We begin with a breakout discussion where participants are asked to consider barriers to investigating and filing academic incidents. Participants are encouraged to generate as many ideas as possible or dive into a deeper discussion about a single barrier. We then return to the group and unpack what each group discussed. I introduce Felicity Prentice's (2020) work on the interrelationship of emotional issues in reporting contract cheating, which includes dealing with the system, increased workload, developing a professional role and identity, impact on relationships with students, and the norms of academic integrity. We consider whether these ideas can be expanded to encompass barriers to navigating and reporting all AIVs. Afterward, I present high-level data from an all-faculty survey my colleagues and I conducted that examined barriers to navigating academic misconduct and the role of emotion.

I discuss how our data revealed nuances in relation to barriers and the number of years faculty have taught (Scurr et al., 2024). We finish the synchronous session by discussing the participant's emotions when navigating misconduct, and I ask them to consider where they perceive themselves on their teaching journey and whether the statistics align with their experience. This grounds the conversation in institutionally relevant data and encourages participants to view themselves as part of a greater teaching collective. Importantly, this conversation leaves participants feeling like they are not alone and sets the stage for the following synchronous session, which explores techniques and resources for moving forward when faced with awkward situations.

The second synchronous session presents participants with situations in which faculty may find themselves, followed by strategies and guidance to help in those cases. For instance, we discuss how to navigate persistent students who send multiple emails outside of regular working hours, students who will not admit error despite overwhelming evidence of academic misconduct, or stressed students who describe suffering from serious mental health concerns as a result of academic misconduct allegations. Participants are shown institutional resources (e.g., email templates, referral contacts) as well as guidance and advice on navigating these types of conversations, whether in person or via email. Finally, we examine cheating in group work and explore strategies for navigating these situations.

To close the session, and in the spirit of the title, "The Awkward Conversation," I ask participants to speak openly about how the Academic Integrity Office can better support faculty when navigating potential academic misconduct. Participants have provided insightful ideas that have helped the Academic Integrity Office plan effective services.

Course Assessment

The assessment for this course is to create a hypothetical program presentation for faculty. The instructions state:

> You've been invited to a program meeting to discuss academic integrity. Specifically, you have been asked to focus on some barriers to investigating/reporting academic integrity violations and why it is important to maintain academic integrity standards in the program. Finally, you have been asked to share resources to help faculty overcome these barriers. *Your main goal is to create awareness for faculty and leave your audience feeling encouraged, supported, and empowered.*

Participants are asked to identify and discuss at least three barriers to investigating/reporting academic integrity violations, identify and discuss at least two reasons why it is important to maintain academic integrity standards in their program or the college generally, and provide at least one resource that may aid in the investigation process and/or develop an aid that may assist with the faculty-student communication process.

Identifying Plagiarism and Maintaining Academic Honesty

Regardless of school, program, or course type, plagiarism exists, and this course broadly explores the concept. I encourage participants to conceptualize plagiarism beyond citing, referencing, and paraphrasing and consider the complex skills and extensive practice required for students to demonstrate their learning through writing and other mediums. It is important to recognize that preventing plagiarism involves fostering an environment where students feel supported in developing their academic voice. By understanding the underlying reasons for plagiarism, educators can implement more effective strategies to promote originality and integrity in students' work. The course aligns with the following Post-Secondary Teaching Certificate program learning outcomes:

- Work effectively within a teaching role in post-secondary settings that align with the expectations of outcomes-based and/or competency-based education.
- Provide an inclusive and culturally relevant learning environment with learning opportunities that support the success of a diverse student population.
- Reflect on aspects of teaching to support the ongoing development of one's identity as a post-secondary educator.

Course Description

- Participants discuss the concept of plagiarism using current definitions, frameworks, and writing pedagogy. They examine the relationship between culture and plagiarism, and the developmental stages of first-language and second-language writers are explored to contextualize phrases like synonym replacement, patchwriting, and too-close paraphrasing. Key concepts like intentionality and evidence of learning are also considered.

Course and Unit Learning Outcomes

1. Interpret different frameworks, definitions, and related terms associated with plagiarism to understand the complexities of plagiarism:
   - Evaluate the historical context of plagiarism.
   - Appraise frameworks and definitions of plagiarism.
   - Explore terms like textual borrowing, paraphrasing, patchwriting, and intertextuality.
2. Evaluate factors that influence the appropriate and inappropriate use of source material to understand what constitutes plagiarism:
   - Interpret situations where reusing source material is appropriate.
   - Determine writing phases of first language and second language learners.

- Explore how various cultures treat source material.
- Appraise concepts of intentionality and deception.

Synchronous Sessions

We begin by discussing three dominant frameworks in which plagiarism has been studied: plagiarism as a regulatory/policy issue, plagiarism as a moral issue, and plagiarism as an issue of learning to write in academia (Adam, 2016) and encourage participants to reflect on which framework they tend to view plagiarism, regardless of whether that is intentional or unintentional. The conversation reveals the multiple dimensions and perceptions faculty have toward the topic.

We then move into a brief history of plagiarism lesson inspired by Shaw (2019) and Jamieson (2019). The overall lesson highlights that integrating source materials and formatting citations and references has been influenced by academics and imitates the requirements for publishing work professionally. In other words, students must research, cite, and reference because that is what scholars do. This conversation intends not to diminish or degrade the importance of paraphrasing, citing, and referencing but to have participants reflect on the value of assessments that follow a more traditional written format, such as reports, essays, or term papers. I hope that by reflecting on our perception of plagiarism, in conjunction with analyzing the value and intentions of our assessments, we may begin to reconceptualize assessment design and move away from a *that's just how things have always been done* mentality.

For one of the breakout room activities, I ask participants to work with their classmates to define plagiarism in four minutes. Participants return, often playfully frustrated with the lack of time they were given; we then unpack what each group came up with. Definitions vary widely, from ideas like taking another's ideas or words to more complex concepts that integrate intentionality and possible factors of cultural relativism. Pecorari (2019) defines plagiarism as follows:

> Plagiarism is a form of academic misconduct involving a writing process which results in an inappropriate relationship between the new text and its source(s). Specifically, plagiarism involves a writer re-using propositional content and/or form of expression from another text without adequately signaling that relationship to the reader and an intention to mislead the reader about that relationship. (p.17)

I offer this definition as it is notably limited to written plagiarism and provides a departure point for a broader conversation about plagiarism in other contexts. I also appreciate the notion of a reader-writer relationship and intentionality.

The next topic we examine is whether culture influences student plagiarism. The conversation begins by discussing table manners and eating etiquette and how certain manners are implicitly expected depending on the context and location. The analogy is applied to writing conventions and standards, as universal expectations do not exist for either. In the Westernized, outcomes-based education system in which I function, paraphrasing, citing, and referencing are expected and copying information from sources without attribution is considered plagiarism. Yet, our students come from a multitude of backgrounds and varying educational systems (Handa & Power, 2005). Our writing conventions are neither better nor worse than others; they are simply different. I do make explicit that this does not mean we should apologize for our standards. Instead, I encourage faculty to consider assessment and rubric design. If citing and referencing, and other elements like paraphrasing, analysis, and evaluation are crucial to the learning outcome, then this should be reflected in the rubric, and if it is done poorly or is simply non-existent in the submission, we should grade the assessment accordingly, even if that means issuing a failing grade. I discourage viewing students' failure to follow the writing standards of our institution squarely within a moral and ethical framework that requires filing an academic integrity offence. I suggest the student would benefit from learning more about our institution's expectations for writing and citing source materials to improve their work.

- We end the session with a discussion on the impacts of multilingualism on learners and various types of *plagiaries* that may result, such as patchwriting, too-close paraphrasing, and careless source attribution. We discuss Jamieson and Howard's (2019) "Rethinking the relationship between plagiarism and academic integrity" and debate whether the term plagiarism even belongs in the academic integrity discourse.

Course Assessment

The assessment for this course requires participants to create a Demystifying Plagiarism student resource: they can make it course or program-specific or keep it general. The overarching goal of this assessment is to create a resource for students that may help alleviate potential anxieties and uncertainties they may have about plagiarism. It consists of three parts; first, participants must provide a clear, non-jargonistic explanation of plagiarism. Second, they must explain why submitting independent and authentic work is important. Here, I encourage participants to situate the conversation in an outcomes-based education framework, requiring students to demonstrate assessment and course outcomes independently. Lastly, they must create a connection to how these transfer to the workplace/industry. This can be an example of how someone cheated or plagiarized in industry or by showcasing the value of independent and authentic work in industry, and I encourage participants to focus on an industry related to the program they teach in.

An Inherent Conflict? Academic Integrity and Technology

When I first began conceptualizing this course, I could not have foreseen its relevance to our educational context in the wake of generative artificial intelligence (GenAI). The overall aim of this course is twofold. The first goal is to promote a critical understanding of the strengths and weaknesses of technology tools to assess potential misconduct without going into detail about all tools. The second goal is to consider technologies our students might use to undermine assessment outcomes and situate them within a cognitive offloading framework to consider assessment (re)design. The course aligns with the following Post-Secondary Teaching Certificate program learning outcomes:

- Work effectively within a teaching role in post-secondary settings that align with the expectations of outcomes-based and/or competency-based education.
- Provide evidence-based, student-centered teaching that is sensitive to student experiences, values, beliefs, and needs.
- Reflect on aspects of teaching to support the ongoing development of one's identity as a post-secondary educator.

Course Description

Has technology in the classroom precipitated an "arms race?" This course examines some benefits and limitations of various technologies and explores strategies for balancing our use of them in supporting potential misconduct investigations. We will explore technology and different tools from a cognitive offloading perspective and begin to strategize how we can reconcile some of the conflicts we, as educators, may find ourselves in, especially if technologies are being used that undermine our assessment expectations.

Course and Unit Learning Outcomes

1. Interpret the nexus of technology and academic integrity and how the two may clash and complement one another to see the trends within an academic integrity framework:
    - Explore generally how technology has both benefits and limitations within higher education.
    - Explore how generative artificial intelligence works and how it may impact academic integrity.
2. Examine potential benefits and limitations of the technologies we use to navigate potential Academic Integrity Violations:
    - Examine how similarity-checking software works and its strengths and limitations when identifying possible misconduct.

- Understand how technology tools, when viewed from a cognitive offloading perspective, impact what is considered acceptable versus unacceptable within the context of academic integrity.

Synchronous Sessions

We begin with a thought exercise where I ask participants whether they typically view technology as a vehicle to promote learning or an antidote to a threat. While there is no right or wrong answer, the conversation generates multiple perspectives, often ones other participants have not considered.

We then look at two technologies for monitoring or checking for possible misconduct: similarity-checking and proctoring software. We do not go into granular detail as our institution offers separate workshops on these topics; we review them from a high-level perspective, consider their strengths and weaknesses, and determine when these tools might prove advantageous.

Next, we turn our attention to artificial intelligence, a topic integrated into this course before GenAI took higher education by storm in late 2022. Initially, this course focussed mainly on predictive artificial intelligence, like grammar and proofreading software, as well as paraphrasing and synonym-replacement software. While we do touch on this, our focus has shifted to mainly consider GenAI.

We finish the first synchronous session with a story from Lars-Erik Nilsson's (2016) work in which he talks about two Swedish students in the early 1900s who developed a small book of formularies (a collection of formulas for the compounding and testing of medication). When the institute discovered these formularies, the students were expelled since, after all, who would create such a small booklet of information without nefarious intent? A few years later, similar formularies were published, sold, and used in industries like labs and pharmacies. Once these formularies became normalized, they were introduced to university students as aids. Students could draw on the formularies, so they did not need to memorize the various formulas, freeing the cognitive load for other, more important mental tasks. As a result, the threat these formularies were once considered to bring was neutralized by integration into the higher education system. I ask participants to percolate on this story for the week, as it forms the basis of our discussion during the second synchronous session.

The second synchronous session begins by showing participants some of the more advanced technologies students have been known to use to cheat. We look at text message calculators, spy glasses, earpieces, and smartwatches. We follow this with a conversation about establishing classroom protocols and best practices to mitigate the use of such technologies.

We then revisit the previous week's story about the formularies and consider the concept of cognitive offloading, described as "The use of physical action to alter the information processing requirements of a task to reduce cognitive demand" (Risko & Gilbert, 2016, p. 677). This is an apt bridge to link academic integrity and assessment design. I provide examples of cognitive offloading tools, like pen and paper, the calculator, auto-paraphrasing software, translation tools, and GenAI, and explain that cognitive offloading tools, in certain contexts, are legitimate tools for the job while in other contexts, they constitute cheating (Dawson, 2021). This provides an opportunity to discuss assessment outcomes and consider how various tools can impede or enhance these outcomes depending on what we are trying to measure. We explore the concepts of automaticity, or doing things via an automatic response, thus allowing for higher cognition (Moors & De Houwer, 2006), authentic assessment, which is an assessment that represents the real world of the discipline (Villarroel et al., 2017), evaluative judgment, which is the capability to make decisions about the quality of work of oneself and others (Tai et al., 2018), and reverse scaffolding.

Once these concepts are discussed, I propose that not only does the intention of the assessment (what it is we are trying to measure) influence assessment design, but considering how students might cheat (the threat to assessment outcomes) can collectively influence assessment design. I then propose four filters that can help determine a course of action for current assessments that may be under threat, which can help to neutralize that threat. The four filters are:

1. *Abandon* to Neutralize asks: Should I abandon or stop using the assessment?
2. *Monitor* to Neutralize asks: Should I monitor the assessment taking place more closely?
3. *Enhance* to Neutralize asks: Should I enhance or improve the assessment?
4. *Adopt* to Neutralize asks: Should I adopt the cognitive offloading tool and incorporate its use into the assessment?

While these filters are still in their infancy in terms of development, they nevertheless serve as a fine thought exercise for this course and encourage participants to consider the value of current assessments, how those assessments may come under threat, and how they might consider assessment redesign.

Course Assessment

The assessment for this course is a Technology Overview: Student Resource. Participants are asked to create a student resource for any technology they use in class. For instance, they can provide a resource on proctoring software used for their midterm or final exam. Alternatively, participants can create a resource on a video creation platform or something entirely different. The overarching goal of the submission is for participants to create a resource that they can apply directly to their assessments and help students become familiar with the technology. The assessment must have an explanation of what the technology is and why it is used, an explanation of the limits of the technology, an explanation of how the technology is not intended to be used, an explanation of how faculty use it to help maintain authentic work and ensure academic integrity (if applicable), and a minimum of three tips or pieces of information that can help students use the technology to ensure an authentic submission.

Contract Cheating in Higher Education

The final course that is offered explores the world of contract cheating in higher education and does so from multiple angles. The main goal of this course is to familiarize participants with the raw and indisputable fact that contract cheating exists. Moreover, it aims to showcase that students participating in contract cheating may be perpetrators but also victims. The intent is to strike a balanced appreciation of the struggles students face that may lead to contract cheating in the face of ruthless and aggressive marketing tactics in conjunction with our institution's no-tolerance stance toward those who do it. This can only happen when we do our best to get in front of and educate our students about the dangers it poses to them, our institutions, and higher education generally. The course aligns with the following Post-Secondary Teaching Certificate program learning outcomes:

- Provide evidence-based, student-centered teaching that is sensitive to student experiences, values, beliefs, and needs.
- Reflect on aspects of teaching to support the ongoing development of one's identity as a post-secondary educator.

Course Description

This course takes an in-depth look at the global phenomenon of contract cheating, an issue many educators consider threatening to higher education. Participants look at the various types of contract cheating services and practices and their aggressive advertising strategies. They also explore the signs of contracted work and learn how to combat contract cheating.

Course and Unit Learning Outcomes

1. Determine what constitutes contract cheating in its various forms to proactively mitigate it:
    - Determine the definition of contract cheating.
    - Appraise various forms of contract cheating services.
2. Explore the marketing and advertising strategies used by contractors to understand how students may experience this:
    - Evaluate how contract cheaters advertise and market their services.
    - Explore how students may experience/perceive contract cheating sales tactics and opportunities.
3. Evaluate signs of contracted work to navigate contract cheating cases:
    - Explore various signs of contracted work.
    - Identify faculty supports to help navigate and combat contract cheating.

Synchronous Sessions

The session starts with a game called *Is This Contract Cheating?* whereby faculty are presented with various situations, and we discuss whether it constitutes contract cheating. It is always fascinating to encounter new faculty who have not heard of or encountered contract cheating; I feel both surprised and admittedly sad for them as this reality comes crashing down on them. This is followed by examining various types of contract cheating services (e.g., third-party file-sharing sites, pre-written essay banks, and freelancers).

After establishing what contract cheating is and what forms it can take, we explore the aggressive marketing and advertising techniques contractors use. I show images of various messages colleagues, students, and I have received from contractors and how they promote their services. I also include an online conversation I had with a contract cheating company where I posed as a desperate student; I asked explicit questions, to which I received reassurance that they have only *the best PhDs* and that *my data was safe with them*. When I said bluntly that I heard this was contract cheating, the staff member assured me that *they have millions of daily visitors*. When I asked if they would finish the rest of my degree, they commented, "Yes, sure, please upload your requirements." It is important to highlight this to participants since it demonstrates how easily our students can succumb to such tactics, especially when they feel desperate, and the pressure is on them to succeed. It is also important so that we can act as a kind of shield for our students by proactively communicating to them that we, as an institution, know all about the industry and are here to support them.

Our next synchronous session begins by examining multiple approaches to combating contract cheating, which includes educating faculty, developing student awareness, considering assessment design, and looking for signs of contracted work, with concepts adapted from Lancaster and Glendinning's (2019) webinar. To enhance assessment design, I point participants to the many resources and educational opportunities our Teaching and Learning department offers.

The remaining time is spent exploring signs of contracted work. We discuss uncharacteristically strong submissions relative to previous submissions, meta-data and title page inconsistencies, odd and overly robust or unrelated bibliographies, and the relevance of IP addresses. I emphasize that all evidence must be qualified and contextualized, and not all evidence is sound on its own. I show participants our *Process Guide*, a resource I co-developed in 2019 to help support staff with meeting students to discuss submissions that raise academic misconduct concerns. The guide offers structure to the conversation and general questions faculty can ask students about the authenticity of their submissions, along with a four-point scoring rubric seen in Fig. 5.2.

Please only ask questions relevant to the assignment in question.

Please score student answers on the following range:

Student demonstrated excellent knowledge of the question.

Student could answer the question satisfactorily.

Student demonstrated very limited knowledge of the question.

Student could not answer the question.

Question	Student could not answer the question.	Student demonstrated very limited knowledge of the question.	Student could answer the question satisfactorily.	Student demonstrated excellent knowledge of the question.
1. Can you please tell me what your assignment is about? Example follow-up questions: - What is the main topic? - What did you argue/discuss/highlight? - What was the focus of the assignment?				
2. What was your favourite part of the assignment, and what was the most difficult part of the assignment?	☐	☐	☐	☐
3. Can you please explain how you went about writing this assignment? (i.e., try and have the student speak about their process & timeline)	☐	☐	☐	☐
4. If there is a reoccurring term or phrase, ask the student to explain what it means. I.e., You use the term "distillation" throughout your assignment; please explain what this term means. For coding/programing assignments, ask the student to explain particular functions.	☐	☐	☐	☐
5. Provide the student with their reference list; select a source and ask how they got that source. (If applicable)	☐	☐	☐	☐
6. Did you make use of technological tools in writing the assignment (i.e., ChatGPT or Microsoft Co-pilot, etc.)? (If applicable)	☐	☐	☐	☐
7. Additional questions:	☐	☐	☐	☐
8. Overall, did the student demonstrate course outcomes for the particular assessment?	☐	☐	☐	☐
Probability of Misconduct (Given the score, is it more likely than not the student submitted work completed by someone else?)	If most answers fall within these two columns, it is more likely the work was completed by someone else		If most answers fall within these two columns, it is more likely it is the student's original work	

Fig. 5.2 Faculty Question Guide for Suspected Misconduct. *Note.* From Conestoga College, Academic Integrity Office

Course Assessment

The assessment for this course consists of a *Contract Cheating Awareness Piece*, and the overarching goal of the submission is to create a resource for students so they can learn the dangers of contract cheating and know that faculty are equipped with an understanding of how to spot it. The *Contract Cheating Awareness Piece* should have a definition or explanation of contract cheating, which may be done in the participant's own words, and at least one example of a scenario that would be defined as contract cheating. Additionally, the participant will provide an overview or explanation of the dangers of contract cheating and at least one resource that can help students avoid contract cheating or academic integrity violations generally.

If it Feels Good, Do it

Each institution is different and, therefore, has different needs. While these five courses currently suit our institution, they may not be the desired focus for others. Moreover, these courses are naturally organic and change over time. Faculty are the backbone of our teaching and, therefore, must be supported in a robust and relevant way. Academic integrity education offers a valuable and fulsome lens through which we can view teaching and learning. As a result, faculty education on academic integrity is key to the successful implementation of sound teaching and learning practices and offers rich, nuanced, and relevant insights and perspectives on teaching in higher education. As a result, I encourage all institutions to consider developing robust and consistent faculty education leveraging an academic integrity perspective. The institution will benefit, faculty will benefit, and most importantly, students will benefit.

References

Adam, L. (2016). Student perspectives on plagiarism. In T. Bretag (Ed.), *Handbook of academic integrity* (pp. 519–535). Springer.

Assiniboine College (n.d.). *Academic Integrity for Faculty*. https://assiniboine.libguides.com/ai-faculty#s-lg-box-16368504

Bertram Gallant, T. (2008). Academic integrity in the twenty-first century: A teaching and learning imperative. *ASHE Higher Education Report, 33*(5), 1–143.

Bertram Gallant, T. (2017). Academic integrity as a teaching & learning issue: From theory to practice. *Theory to Practice, 56*(2), 88–94. https://doi.org/10.1080/00405841.2017.1308173

Bertram Gallant, T. (2022, February 11). "When a thousand flowers bloom": The academic integrity rating system (AIRS). *International Center for Academic Integrity*. https://academicintegrity.org/resources/blog/96-2022/february-2022/345-when-a-thousand-flowers-bloom-the-academic-integrity-rating-system-airs

Bertram Gallant, T., & Drinan, P. (2008). Toward a model of academic integrity institutionalization: Informing practice in postsecondary education. *Canadian Journal of Higher Education, 38*(2), 25–43.

Conestoga College. (n.d.). *Micro-credentials*. Continuing Education. https://continuing-education.conestogac.on.ca/microcredentials

Dawson, P. (2021). *Defending assessment security in a digital world: Preventing e-cheating and supporting academic integrity in higher education*. Routledge.

Egan, A. (2018). *Improving academic integrity through assessment design*. Dublin City University. https://www.dcu.ie/sites/default/files/2020-10/academic-integrity-literature-review_0.pdf

Gagne, M. (Ed.). (2014). *The oxford handbook of work engagement, motivation, and self-determination theory*. Oxford University Press.

Handa, N., & Power, C. (2005). Land and discover! A case study investigating the cultural context of plagiarism. *Journal of University Teaching & Learning Practice, 2*(3). https://doi.org/10.53761/1.2.3.8

Jamieson, S. (2019). Is it plagiarism or patchwriting? Toward a nuanced definition. In T. Bretag (Ed.), *Handbook of academic integrity* (pp. 508–509). Springer.

Jamieson, S., & Howard, R. M. (2019). Rethinking the relationship between plagiarism and academic integrity. *International Journal of Technologies in Higher Education, 16*(2), 69–85. https://doi.org/10.18162/ritpu-2019-v16n2-07

Kezar, A., & Bernstein-Sierra, S. (2016). Commercialization of higher education. In T. Bretag (Ed.), *Handbook of academic integrity* (pp. 325–346). Springer.

Lancaster, T., & Glendinning, I. (2019). *Reducing contract cheating – more international opportunities for action* (Webinar). https://www.youtube.com/watch?v=UQoljtfNAeY

McCabe, D. L., & Pavela, G. (2004). Ten (Updated) principles of academic integrity: How faculty can foster student honesty. In Change (New Rochelle, N.Y.) (Vol. 36, Number 3, pp. 10–15). *Taylor & Francis Group*. https://doi.org/10.1080/00091380409605574

Moors, A., & De Houwer, J. (2006). Automaticity: A theoretical and conceptual analysis. *Psychological Bulletin, 132*(2), 297–326. https://doi.org/10.1037/0033-2909.132.2.297

Nilsson, L. E. (2016). Technology as a double-edged sword: A promise yet to be fulfilled or a vehicle for cheating? In T. Bretag (Ed.), *Handbook of academic integrity*. Springer.

Pecorari, D. (2019). Can plagiarism be defined? In D. Pecorari & P. Shaw (Eds.), *Student plagiarism in higher education* (pp. 12–27). Routledge.

Prentice, F. (2020). Impediments to reporting contract cheating: Exploring the role of emotions. In T. Bretag (Ed.), *A research agenda for academic integrity*. Edward Elgar Publishing.

Risko, E. F., & Gilbert, S. J. (2016). Cognitive offloading. *Trends in Cognitive Sciences, 20*(9), 676–688. https://doi.org/10.1016/j.tics.2016.07.002

Scurr, C., Mangahis, C., & Mandal, J. (2024). *Toward a nuanced understanding: Faculty barriers to navigating academic integrity violations*. [Manuscript submitted for publication]. Conestoga College.

Shaw, P. (2019). Are we making our students plagiarize? In D. Pecorari & P. Shaw (Eds.), *Student plagiarism in higher education* (pp. 125–135). Routledge.

Tai, J., Ajjawi, R., Boud, D., Dawson, P., & Panadero, E. (2018). Developing evaluative judgement: Enabling students to make decisions about the quality of work. *Higher Education, 76*(3), 467–481. https://doi.org/10.1007/s10734-017-0220-3

Villarroel, V., Bloxham, S., Bruna, D., Bruna, C., & Herrera-Seda, C. (2017). Authentic assessment: Creating a blueprint for course design. *Assessment & Evaluation in Higher Education, 43*(5), 840–854. https://doi.org/10.1080/02602938.2017.1412396

Wade, S. L. (2017). *Self-determination theory (SDT): Perspective, applications and impact*. Nova Science Publishers.

Dr. Cory Scurr works at Conestoga College as Associate Director of the Academic Integrity Office. In this role, he leads a team of Academic Integrity Specialists and Associates and supports faculty, students, and administrators through numerous academic integrity initiatives, consultations, student outreach, and resource development. He is also responsible for developing and delivering Conestoga's Micro-Credential on Academic Integrity. Dr. Scurr holds a Ph.D. from Wilfrid Laurier University and has worked in higher education for over a decade where he has developed and delivered courses in both College and University settings. Dr. Scurr has presented a number of conference presentations provincially, nationally, and internationally on various academic integrity-related topics. Finally, he is the current Chair of the Academic Integrity Council of Ontario (AICO).

Chapter 6
Building Relational Culture in Polytechnic Education Through Restorative Approaches to Academic Integrity

Alana Fukushima and Linden Couteret

Abstract The inherent relationships between polytechnic institutions and related industries underscore the importance of community, aligning the primary function of polytechnic education with principles of restorative practices, where strong relationships and social connections within communities are fundamental. Through this discussion, we explore how restorative practices resonate deeply with the holistic, engaged, and active participation central to polytechnic education by valuing active accountability and collaboration to prevent and repair harm. These practices also connect to applied experiential learning by emphasizing active accountability and participatory learning. Furthermore, the polytechnic emphasis on innovation and innovation-enabling skills aligns well with restorative practices, which require consideration of the specific harms and obligations of each unique situation. In this chapter, we will: a) discuss the nature of restorative practices, b) explore how restorative processes support the goals of polytechnic education, c) examine considerations for implementing restorative practices, d) present practical applications of restorative practices in polytechnic education, and e) tackle the challenges of implementation while outlining opportunities for individuals to adapt these practices.

Keywords Academic integrity · Polytechnic · Restorative practices · Vocational training · Integrity principles · Community · Accountability

---

A. Fukushima (✉)
Northern Alberta Institute of Technology (NAIT), Edmonton, AB, Canada
e-mail: alanaf@nait.ca

L. Couteret
University of Alberta, Edmonton, AB, Canada
e-mail: couteret@ualberta.ca

L. Vogt, B. M. Stoesz (eds.), *Academic Integrity in Vocational and Polytechnic Education*, Ethics and Integrity in Educational Contexts 8,
https://doi.org/10.1007/978-3-031-90091-4_6

> "To a considerable degree, restorative justice is a rebellious act of creative imagination." (Llewellyn et al., 2013).

We write this chapter to imagine how a polytechnic institution can embrace the tenets of restorative practices to strengthen the principles and aims of polytechnic education. Throughout this work, we describe how restorative practices have the power to build a thriving culture of academic integrity and promote a relational mindset where students and staff value their roles in the learning community. Our approach is guided by our experiences working with restorative practices, which inform our deep belief in reducing harm through understanding and addressing the root causes of academic integrity breaches.

Polytechnics Canada (2023) states that polytechnics are "post-secondary institutions that offer advanced technical education," defining polytechnic education according to the following traits: applied and hands-on learning, industry-driven education and curricula, flexible life-long learning, and focus on innovation and applied research. When a student obstructs the accurate assessment of knowledge, skills, and abilities, it undermines the key outcomes and traits of polytechnic education.

In this chapter, we discuss how restorative practices support the aims of polytechnic education, offer strategies for building a culture of academic integrity through restorative practices, and explore practical ways to implement restorative practices at the individual and institutional levels. We believe that restorative practices have much to offer polytechnic education, particularly regarding academic integrity, and we are grateful for the opportunity to share our thoughts.

Positionality & Reflection

It is important to reflect on and share our own position in relation to the topics being discussed. Both authors work in the field of student conduct and have experience using restorative practices at polytechnic institutions. We were both born and raised on Treaty territory in what is known as Alberta, Canada (Treaties 6 & 8) and currently reside in Treaty 6 territory in/around Edmonton, also known as Amiskwaciwâskahikan. We both come from families where English is the primary language; we are both white and descend from European heritage. We have completed the majority of our education in the Canadian public school system and at public universities. We each have education and experience in restorative justice (RJ), including a combination of job-related training, practical workshops, and certificate courses, as well as experience in roles facilitating and implementing restorative processes.

Introductions to restorative practices often refer to them as coming from or being influenced by Indigenous practices and paradigms without critically reflecting on the cooption (one could even say colonization) of Indigenous practices into a settler worldview. Littewolf et al. (2020) provide an example in the metaphor of restorative justice as a bridge to Indigenous communities and worldviews while pointing out that it is false to consider the current restorative justice movement to be rooted in

Indigenous ways. Valandra (2020) notes that "acknowledging RJ's Indigenous antecedents [...] is not the same as addressing or undoing settler colonialism's First Harm" (p. 344). Restorative practices may borrow from or even replicate Indigenous processes and protocols but do so within settler worldviews and generally fail to prioritize repairing the deeper, systemic harms that underlie the foundations of both Canada and the United States. While this chapter considers restorative practices in the polytechnic environment a valuable approach, particularly in relation to a culture of academic integrity, it is also necessary to reckon with the current and historical contexts of restorative practices in Canada.

The colonization of people and land is often acknowledged, but it is also important to acknowledge the colonization of knowledge and culture. In discussing how we can re-center relationships, our interconnectedness, and community, which are prevalent values in Indigenous paradigms (Absolon, 2010; Gladue, 2020), we recognize our privilege and responsibility for allyship and advocacy in the decolonization of this work.

Restorative Practices

There are several definitions of restorative practices and restorative justice. We have adopted the International Institute for Restorative Practices (IIRP) approach, where restorative practices are considered the umbrella under which restorative justice falls. Fania Davis describes restorative justice in The Little Book of Race and Restorative Justice (2019) as:

> ...profoundly relational and emphasizes bringing together everyone affected by wrongdoing to address needs and responsibilities and to heal the harm to relationships and community, to the degree possible. While often mistakenly considered only a reactive response to harm, restorative justice is also a proactive relational strategy to create a culture of connectivity where all members of a community thrive and feel valued. (p. 19)

Restorative practices look at strengthening relationships between individuals and social connections within communities (IIRP, 2018). In higher education, restorative practices are commonly discussed as an approach to responding after misconduct (Kara & MacAlister, 2010; Karp, 2019; Zehr, 2003), but restorative practices are more than what happens after harm. They encompass small actions and day-to-day interactions that foster relationships before harm has occurred, including community building, strengthening relationships, and improving school climate and culture (Hickman, 2022).

Costello et al. (2019) provide examples of restorative practices in education, including:

- Having small impromptu conversations to encourage and support desired behaviour.
- Using affective statements when discussing behaviour, which focuses on positive or negative impacts rather than morality or compliance, to build relationships and encourage desired behaviour.

- Addressing concerning behaviour before it escalates and must be tended to more formally.
- Asking affective or restorative questions when misconduct occurs, such as what a student was thinking about or feeling when they did something, who has been affected and how, and how they can try to repair harm.
- Using specific protocols for addressing harm, such as talking circles.

The underlying principles of restorative justice include the importance of shared values that prioritize connection, mutual understanding, and relationships, as well as active accountability when harm has occurred, a focus on building relationships and community, engagement in decision-making, and learning through participation in practices and processes. Restorative justice asks us to go beyond individual instances of harm and reflect on our underlying principles, ideals and goals (Llewellyn et al., 2013).

For individuals who do not have control over institutional processes, proactive restorative practices can be implemented on an individual level in everyday interactions by making intentional choices that prioritize relationship and community and selecting supportive pedagogical approaches. For example, Zehr (2003) further emphasizes values in restorative justice (and restorative practices in general), stating that "one of the greatest attributes is the way it encourages us to explore our values together." In relation to academic integrity, restorative practices encourage us to focus on the six fundamental values from the International Center for Academic Integrity (2021):

- **Honesty**: being truthful, giving credit where it is due, and considering multiple perspectives
- **Trust**: reciprocal trust demonstrated by stating expectations, promoting others to act with genuineness, and seeking mutual understanding
- **Fairness**: demonstrated by equitable treatment, consistent application of policies and rules, and objectivity
- **Respect**: active listening, empathy, treating others and their opinions as valid, and being willing to receive feedback
- **Responsibility** at the individual and community level
- **Courage** to hold true to one's values even in stressful or fearful circumstances

Differentiating Restorative Practices

To understand how restorative practices may differ from or add to traditional views of academic integrity, it is important to first define institutional approaches to academic integrity that have long been applied in education. Bertram Gallant (2011) identifies two as follows:

- **Rule compliance approaches** have dominated higher education and focus on enforcing rules, often with limited explanations for their purpose and intention,

with follow-up discipline for students who violate the rules. In application, these are typically legalistic and/or adversarial in nature and seek to change behaviour by increasing the risk, detection, and cost of misconduct.
- **Integrity approaches** have gained prominence and focus on what students should do rather than what they should not do. These approaches seek to guide students' development in making ethical choices and acting with integrity by using incidents of misconduct as teaching opportunities.

Rule compliance approaches assume that academic integrity is best upheld through deterrent penalties for misconduct and that punitive responses dissuade students from engaging in future breaches of academic misconduct. Integrity approaches also center considerations of integrity and misconduct on the individual student's actions in relation to a set of rules established by the institution. However, both responses overlook the harm that students, instructors, and other interest holders experience through academic misconduct, and there is no opportunity to address or repair this harm. They also do not focus on strengthening the student's relationship with the instructor or the institution. Bertram-Gallant discussed the need to understand the cultural factors that lead to unethical behaviour in order to address integrity as a systemic issue. There is a need for institutional leaders to understand the "processes that motivate individual...decisions" through listening to people talk about their decisions and actions (Bazerman & Tenbrunsel, 2011, pp. 160–161, as cited in Bertram-Gallant, 2024).

Social Discipline Window

Wachtel's (1999) *Social Discipline Window* in Fig. 6.1 helps to visualize how restorative approaches differ from other approaches.

The "to" quadrant represents punitive responses focusing on control rather than discipline. This quadrant would include rule compliance approaches to academic misconduct. By contrast, the "for" window focuses on permissive approaches that offer high amounts of support with little discipline or control. An integrity approach to academic integrity may fall into this quadrant if it does not also include meaningful ways to address academic misconduct. The "not" quadrant represents approaches that are low on support and control, for example, when an instructor chooses not to address a concern of academic misconduct at all. Restorative approaches balance a high level of control with a high level of support, encouragement, and nurturing, resulting in a prioritization of relationships, clear expectations, and shared responsibility.

Bringing restorative practices into the realm of academic integrity seeks to balance support and control by understanding the drivers that foster integrity, acknowledging the impact on relationships within the institutional community, and exploring and addressing the harms caused by academic integrity breaches.

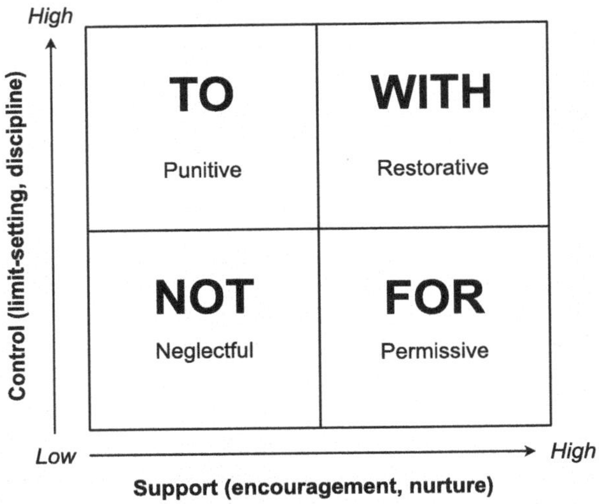

Fig. 6.1 Social Discipline Window. *Note*. Wachtel, T. (1999, February 16–18). *Restorative justice in everyday life: Beyond the formal ritual* [Paper presentation]. Reshaping Australian Institutions Conference: Restorative Justice and Civil Society, The Australian National University, Canberra, Australia. https://www.iirp.edu/eforum-archive/4221-restorative-justice-in-everyday-life-beyond-the-formal-ritual. Copyright International Institute for Restorative Practices. Used with permission

Connecting Polytechnic Education, Restorative Practices, and Academic Integrity

As previously mentioned, Polytechnics Canada (2023) defines the traits of polytechnic education as applied and hands-on learning, industry-driven education and curricula, flexible life-long learning, and a focus on innovation and applied research. In this section, we describe the link between these traits and the nature of restorative practice.

Applied Hands-On Learning

Restorative practices and polytechnic education both emphasize holistic, engaged, active participation and applicable approaches that value our relationship with one another and the ways we are interconnected within the community. Restorative practices create space for students to be actively involved in the realm of academic integrity. The first is in creating institutional culture, which is a keystone of both academic integrity and restorative practices. Hickman (2022) reminds us that the proactive core of restorative practices is building a connected community culture with shared principles, such as the values of academic integrity discussed above. Bertram Gallant and Drinan (2008) discuss the importance of a collaborative approach between academic administrators, student affairs staff, and faculty when building a culture of academic integrity. They also speak to the value of actively involving students in promoting academic integrity. Engaging students in establishing and supporting this culture creates space for valuable learning that can carry forward to other institutional cultures they encounter during their careers.

Restorative practices fit the mandate of applied experiential learning by focusing on active accountability and participatory learning. Dominant approaches to academic misconduct typically require little active involvement from the student, particularly in rule compliance approaches, where a student responds to an allegation and then waits to (potentially) receive a sanction. There is little learning through the process itself, even if the sanction is educational in nature. By contrast, a restorative approach involves multiple skills which are valuable in the workplace: engaging in a difficult conversation, reflecting on actions and their impacts, and collaborative problem-solving. Restorative approaches also create space for larger discussions of accountability and professionalism, managing stress under pressure, and balancing working independently and being part of an interdependent team.

Industry-Driven Education and Curricula, and Flexible Life-long Learning

Like life-long learning, restorative practices are long-term (Restorative Practice Consortium, 2017). Polytechnic programs are often "modular, with stackable credentials that allow students to move between education and the labour market" (Polytechnics Canada, 2023). Life-long learning in polytechnic education means a student may have a relationship with an instructor that extends beyond the course and into future credentials or industry connections. Instructors in polytechnic education are often strongly connected with industry partners, resulting in relational and reputational impacts on academic quality and integrity that extend outside of the educational institution. A student may find their instructor an important reference, a future coworker, or even an eventual employer. Restorative practices create opportunities for students to meaningfully engage and contribute to the integrity of their profession through prioritizing relationships. When a breach of academic integrity is suspected, a restorative process can preserve and repair the reputation between students, instructors, and industry. A meaningful discussion between a student and an instructor on how to (re)build trust after an incident of academic misconduct can be a much more powerful tool in repairing that relationship than a series of forms and an assigned sanction.

The skills strengthened through restorative practices align with Polytechnics Canada's perspective that "report after report on the future of work speaks to the skills that set humans apart, including dynamic problem-solving, teamwork and adaptability" (Polytechnics Canada, 2023). Sternberg (2013) states, "93 percent of employers surveyed said that 'a demonstrated capacity to think critically, communicate clearly, and solve complex problems is more important than [a candidate's] undergraduate major'" (p. 1). Based on the expressed needs from industry and our experiences working with staff and students, we believe that Restorative Practices strengthen the skills desired by employers and contribute to innovation and economic growth in the real world (Benjamin et al., 2013; Levy & Murname, 2004, as cited in Lenci, 2018).

Innovation and Applied Research

Finally, the polytechnic focus on innovation and innovation-enabling skills connects to the restorative approach to craft unique responses to academic misconduct that address the specific harms and obligations of each situation. Whereas traditional responses to academic misconduct often include a standard set of sanctions (e.g., grade reduction of x percent, having to retake an assessment, reviewing a module or course on academic integrity), these are not always rationally connected to the specifics of a situation. Restorative practices focus on meaningfully engaging the student in reparations and building effective communication strategies that support further professional development. This requires listening to understand the factors that lead to the misconduct and meeting the needs that resulted from the misconduct itself. Each response is context-specific, and an innovative skillset is required to understand what has happened, the context under which it happened, what it has meant since it occurred, and how to make it right.

The Conference Board of Canada's Innovation Skills Profile 2.0 outlines the skills, attitudes, and behaviours needed to contribute to innovation in the workplace, which includes skills in creativity, problem-solving and continuous improvement, risk assessment and risk-taking, relationship building and community skills, as well as implementation skills (The Conference Board of Canada, 2014). These are directly connected to Hopkins' (2002) overview of the skills required to implement restorative practices in a community setting, such as a post-secondary institution, which are:

- The ability to have difficult conversations, along with related skills of active, non-judgemental listening, assertiveness, and non-violent communication.
- A focus on demonstrating empathy and building rapport, valuing others explicitly, acknowledging and appreciating diversity, and connecting across differences.
- Demonstrated emotional literacy, such as the ability to understand and manage anger and the ability to develop and maintain self-esteem.
- A desire to constructively challenge oppression and prejudice.

Considerations for Implementing Restorative Practices at a Polytechnic

Zehr (2003) reminds us that restorative justice is not a particular program or blueprint but rather a set of principles and a framework for thinking about wrongdoing. With this in mind, we note that introducing restorative practices does not necessarily mean a complete overhaul of an existing academic integrity system. At the same time, restorative practices cannot be effectively introduced if treated as a checkbox item that can be simply added to a procedure. This section introduces strategies to build restorative practices related to academic integrity.

Foundational to all our suggestions is offering training in restorative practices and intercultural competence and the importance of relational and procedural fairness to all staff involved in responding to academic integrity breaches (Chory, 2023). Attempting to introduce restorative practices without properly preparing staff can lead to ineffective attempts that are not truly restorative and may fail to meet procedural fairness requirements.

Practical Applications

> "Restorative approaches are about more than just repairing relationships when things go wrong, they are also about making and maintaining relationships, and they inform a style of teaching just as much as they do an approach to problem solving and conflict resolution. By giving everyone a voice, considering everyone's thoughts, feelings and needs and believing in people's ability to find solutions to their own problems by working together, young people will develop the language and skills they need to properly engage with their peers" (Hopkins, 2011. para. 1).

Polytechnic instructors and staff can use restorative approaches to contribute to a culture of academic integrity by:

- Prioritizing community and relationships in the classroom through learning students' names, interests, and motivations and sharing with the classroom about who they are, what interests them, and what motivates them (Ellis & Tod, 2009).
- Using active learning strategies that promote engagement and relationship building and make outcomes meaningful to students (Hopkins, 2011).
  - An active learning environment is defined as "one in which students are engaged in the process of building, texting, and defining their mental models" (Michael & Modell, 2003). An example of this is a *think, pair, share* activity, where learners individually reflect on information, pair up and discuss with a peer, and then take turns sharing with the larger group. This active learning strategy promotes student interaction, co-learning, and relationship development (Bickmore, 2011; Ferlazzo, 2020.).

- Engaging in cultural competence training to better understand one's own perspective and the perspectives of others, build empathy across diversity, and create a classroom culture of safety and belonging for all.
- Modeling the values of academic integrity at all levels of the institution, including in the classroom and in providing services for students.
- Demonstrating a commitment to equity through words and actions to create meaningful opportunities for equity-deserving groups and challenge systemic barriers (Eaton, 2022).
- Teaching students what academic integrity looks like and allowing students to learn specific academic integrity acts (such as citations) through practice and making mistakes without being penalized for errors.

Ongoing Discussion

Building a culture of academic integrity involves incorporating regular discussion of academic integrity into the learning environment, including what it looks like in the specific context of that course. This is especially important for activities not commonly included in academic integrity training, such as technical drawings, coding, or creating something in a lab, where instructors should explicitly describe how to demonstrate academic integrity. This is a good practice as it provides clarity for students to understand what is expected of them. For example, an instructor reminding students of an upcoming major exam can:

- Review specific academic integrity expectations, including any requirements (such as how to properly show their environment for an online exam, whether blank scrap paper is permitted, or what students are allowed to bring into the room for an in-person exam).
- Remind students of the values of academic integrity (honesty, trust, fairness, respect, responsibility, and courage) and express commitment to the values, including any work the instructor has done to demonstrate them (for example, explaining that the exam has been updated from recent years to ensure that it can be reasonably done in the required time as a consideration of fairness).
- Ask the students to reflect on why academic integrity is important to the shared community of practice or industry in relation to this assessment and what the harms caused by academic misconduct are. This does not need to be a fulsome or lengthy discussion each time.

We wish to note that the way in which one goes about these conversations is critical. Staff members should not illicit shame or come from a place of judgment, make moralistic statements of 'good' or 'bad,' or expect non-compliance. We communicate with more than just our words; the tone, context, and body language of the conversations are as important as the exact words said. A restorative discussion of academic integrity is one that invites relationship-building and collaboration rather than fear.

Meaningful Involvement in Decisions

Hands-on learning includes having the opportunity to practice real-world tasks and learn through doing. Active accountability can be seen as a form of hands-on learning where students take a direct role in practicing and demonstrating integrity. This can be practiced within dominant models of academic integrity by meaningfully involving students in decisions about how to move forward from academic integrity breaches. Similarly, affected instructors can be included in processes and given space to reflect on and share the impacts of academic misconduct and seek meaningful repair. This approach brings us back to the importance of centring relationships and accountability, which allows us to address the negative emotional impacts of academic misconduct on instructors' workload and well-being, as identified by Schneider (2023), elements that are not typically considered in compliance-focused responses. Finally, this creates space to identify and address the factors that contribute to academic misconduct. As Kaba (2021) discussed, a system that never addresses the why behind a harm will never actually reduce the harm itself.

Involving students and impacted instructors can be informal, such as having an open discussion with each party and providing an opportunity to contribute their thoughts, or it can be formal, using specific restorative practices. Sopcak and Hood (2022) give an example of a formal approach used in a university, where procedures provide an option to divert to a restorative resolution, primarily using a restorative conference to collaboratively determine outcomes. A restorative conference is a structured process with a set speaking order and a standard list of questions to explore what happened, what harm was caused, and what needs to happen to repair the harm.

For meaningful involvement and to promote active accountability, students can be asked to reflect on:

- The factors that contributed to their engaging in academic misconduct

  e.g., *"What were you thinking and feeling at the time? What factors led to you making this decision?"*

- The impacts of the misconduct

  e.g., *"Who has been affected by this? How have they been affected? How does this harm the student, other students, the instructor, the institute?"*

- How they can work to repair harm that has been caused

  e.g., *"What could you do that would help repair (harm)? If you were in my position, what would you think was an appropriate response?"*

- How to rebuild trust and relationships with those affected

  e.g., *"What can you do to build trust with your instructor?"*

Meaningful questions to engage harmed instructors (or other harmed parties) include:

- Asking about the personal impacts of the misconduct that they have experienced

  e.g., "What have you thought and felt about regarding this incident? What has been the biggest impact?"

- What would help repair the harm that has been caused

  e.g., "What do you hope comes out of this sanction? What would be meaningful to you in resolving this?"

- What actions by the student would contribute to rebuilding trust

  e.g., "What could the student do to rebuild your trust?"

As an example of meaningful engagement in practice, consider this incident of academic misconduct during the height of the COVID-19 pandemic. After an academic integrity breach, a student shared the details of the stress that was a factor in their decision-making. The student was still living in their home country due to COVID-19 pandemic restrictions while unexpectedly trying to do online school with insufficient internet infrastructure. They were experiencing considerable stress trying to figure out where they would be living mid-semester as they made plans to relocate to Canada with travel restrictions lifting. The nature of the academic integrity breach resulted in a grade of zero for the paper. At the same time, the institution also responded with support for finding a place to rent in Canada, regular check-in meetings with the student that focused on building a relationship, supporting their academic success, and making connections to resources to assist with the logistics of the move and settlement once they arrived. The student shared that they felt incredibly cared for even though they had made a mistake and committed to never breaching academic integrity again.

Challenges to Implementing Restorative Practices at a Polytechnic

Shifting an institutional culture requires buy-in and collaboration to establish shared values, beliefs, and practices. Kara and MacAlister (2010) note that the key to successfully transitioning towards a restorative approach to academic integrity breaches includes ensuring restorative justice values and principles permeate through the institution. With restorative approaches centring on the concept of relationships, we are required to reflect on the larger question of how we build and value relationships within our systems. Post-secondary structures include hierarchical frameworks in which student participants are, in some regard, viewed as customers (Hersh & Keeling, 2013). In a webinar for Epigeum, Bretag (2019) discusses how institutions play a critical role in establishing the framework for a culture of academic integrity. She asks:

> When you walk around campus, do you say, 'Wow! This university really values honesty, trust, respect, fairness, and responsibility.' Or do you walk around saying, 'All I can see is branding and marketing.' (Epigeum, 2019)

Buy-in can also be influenced by the same relationships to industry that have been highlighted above. Lucien-Bay et al. (2025) explore how the close ties between polytechnic institutions and industry can heighten the concern about reputational risk if student learning is misrepresented or students engage in unethical behaviour. Further discussed is the influence of industry regulatory bodies, which may have specific expectations for handling academic misconduct. Restorative responses to harm may be mistakenly perceived as inherently more lenient or less suitable for major offences, described as soft or lacking the ability to hold people accountable (Vaandering & Reimer, 2019). Paul (2015) encourages restorative practice advocates to avoid dualistic discussion between restorative and retributive approaches and to instead focus on common desires.

Finally, a challenge at all post-secondary institutions is ensuring that staff have the resources and skill set identified in this chapter for engaging in restorative practices. Demonstrating empathy and rapport building, having difficult conversations, and understanding and identifying oppression are skills that require learning, reflection, and practice to develop, as well as time to carry out. It is incorrect to assume that all staff at any educational institution inherently have these skills; In our practice, we see a broad range of abilities. Further, instructors already experience time constraints and a heavy toll from the emotional labour of addressing academic misconduct (Schneider, 2023). Instructors have limited time for marking and addressing academic misconduct allegations while also needing to meet institutional deadlines. Attempts to implement restorative practice without deep support for instructors' time and resources, including training to develop the required skills, are likely to fail to produce meaningful outcomes and may, therefore, lead to rejection of and resistance to restorative approaches.

Conclusion

"To the degree possible, restorative justice seeks healing for all versus a victory for one" (Davis, 2019, p. 27). We have used this chapter to envision how restorative practices serve to strengthen polytechnic education and build a culture of integrity by focusing on relationships. We have brought you along by connecting the tenets of restorative practices to the aims of polytechnic education, providing examples of practical applications and discussing the challenges in doing so. We encourage reflection on how individuals can model the values of restorative practice that we have described to enhance polytechnic education and academic integrity, regardless of their power to change institutional policy and practice.

References

Absolon, K. (2010). Indigenous wholistic theory: A knowledge set for practice. *First Peoples Child & Family Review, 5*(2), 74–87. https://doi.org/10.7202/1068933ar

Bazerman, M. H., & Tenbrunsel, A. E. (2011). *Blind spots: Why we fail to do what's right and what to do about it*. Princeton University Press.

Benjamin, R., Klein, S., Steedle, J., Zahner, D., Elliot, S., & Patterson, J. (2013). *The case for critical thinking skills and performance assessment*. Council for Aid to Education. https://issuu.com/claplus10016/docs/the_case_for_critical_thinking_skil

Bertram Gallant, T. (2011). *Building a culture of academic integrity* (White paper). Magna Publications. https://www.depts.ttu.edu/tlpdc/Resources/Academic_Integrity/files/academicintegrity-magnawhitepaper.pdf

Bertram Gallant, T., & Drinan, P. (2008). Toward a model of academic integrity institutionalization: Informing practice in postsecondary education. *Canadian Journal of Higher Education, 38*(2), 25–24. https://journals.sfu.ca/cjhe/index.php/cjhe/article/view/508/557

Bertram-Gallant, T. (2024). Leveraging institutional integrity for the betterment of education. In S. E. Eaton (Ed.), *Second handbook of academic integrity* (pp. 1889–1904). Springer Nature Switzerland. https://doi.org/10.1007/978-3-031-54144-5_52

Bickmore, K. (2011). *Location, location, location: Restorative (educative) practices in classrooms*. https://www.educ.cam.ac.uk/research/programmes/restorativeapproaches/seminarfour/BickmoreSemianr4.pdf

Chory, R. M. (2023). Fairness matters in higher education: Student classroom justice perceptions and behavioral responses. In *9th International conference on higher education advances (HEAd'23)* (pp. 385–392). https://doi.org/10.4995/HEAd23.2023.16315.

Costello, B., Wachtel, J., & Wachtel, T. (2019). *The restorative practices handbook* (2nd ed.). International Institute for Restorative Practices.

Davis, F. E. (2019). *The little book of restorative justice*. Good Books.

Eaton, S.E. (2022). New priorities for academic integrity: equity, diversity, inclusion, decolonization and Indigenization. *International Journal for Educational Integrity 18*(10). https://doi.org/10.1007/s40979-022-00105-0

Ellis, S., & Tod, J. (2009). *Behaviour for learning: Proactive approaches to behaviour management*. Routledge.

Epigeum. (2019, January 7). *Dr Tracey Bretag: How can institutions create a culture of academic integrity?* (Video). YouTube. https://www.youtube.com/watch?v=JgNwlQ2q49k

Ferlazzo, L. (2020, January 9). *Ways to implement restorative practices in the classroom*. Education Week. https://www.edweek.org/teaching-learning/opinion-ways-to-implement-restorative-practices-in-the-classroom/2020/01

Gladue, K. (2020). Indigenous academic integrity [PDF]. Taylor Institute for Teaching and Learning, University of Calgary. https://taylorinstitute.ucalgary.ca/sites/default/files/Content/Resources/Academic-Integrity/21-TAY-Indigenous-Academic-Integrity.pdf

Hersh, R. H., & Keeling, R. P. (2013). Changing institutional culture to promote assessment of higher learning. (Occasional Paper No. 17). Urbana, IL: University of Illinois and Indiana University, National Institute for Learning Outcomes Assessment (NILOA).

Hickman, K. (2022). *Creating cultural change in education: Implementation science and human capital theory*. IIRP Presidential Paper Series, Number 7. https://www.iirp.edu/images/2023/Creating_Cultural_Change_in_Education.pdf

Hopkins, B. (2002). Restorative justice in schools. *Support for Learning, 17*(3), 144–149. https://doi.org/10.1111/1467-9604.00254

Hopkins, B. (2011). *The restorative classroom: Using restorative approaches to foster effective learning*. Speechmark.

International Center for Academic Integrity. (2021). *The fundamental values of academic integrity* (3rd ed.). International Center for Academic Integrity. https://academicintegrity.org/images/pdfs/20019_ICAI-Fundamental-Values_R12.pdf

International Institute for Restorative Practices (IIRP). (2018). *What is restorative practices?* (Video). https://www.iirp.edu/restorative-practices/explained

Kaba, M. (2021). *We do this 'til we free us: Abolitionist organizing and transforming justice, a system that never addresses the why behind a harm will never actually reduce the harm itself.* Haymarket Books.

Kara, F., & MacAlister, D. (2010). Responding to academic dishonesty in universities: A restorative justice approach. *Contemporary Justice Review, 13*(4), 443–453. https://doi.org/10.1080/10282580.2010.517981

Karp, D. R. (2019). *The little book of restorative justice for colleges and universities: Repairing harm and rebuilding trust in response to student misconduct.* Good Books.

Lenci, S. (2018). *Industry, faculty, and students' perceptions of the opportunities and constraints regarding the implementation of professional skills into technical education programs.* Doctoral thesis. University of Calgary. https://prism.ucalgary.ca; http://hdl.handle.net/1880/106487; https://doi.org/10.11575/PRISM/31775

Levy, F., & Murname, R. J. (2004). Education and the changing job market: An education centered on complex thinking and communication is a graduate's passport to prosperity. *Educational Leadership, 62*(2), 80–83.

Littewolf, E., Armster, M., Paras, C., & Stutzman Amstutz, L. (2020). Burn the bridge. In E. C. Valandra & W. W. Hiksila (Eds.), *Colorizing restorative justice: Voicing our realities* (pp. 325–370). Living Justice Press.

Llewellyn, J. J., Archibald, B. P., Clairmont, D., & Crocker, D. (2013). Imagining success for a restorative approach to justice: Implications for measurement and evaluation. *Dalhousie Law Journal, 36*(2), 281–316.

Lucien-Bay, L., Couteret, L., & Radke, S. (2025). Polytechnic pinch points for academic integrity: Reflections from student conduct officers. In L. Vogt & B. M. Stoesz (Eds.), *Academic integrity in vocational and polytechnic education*. Springer.

Michael, J., & Modell, H. I. (2003). *Active learning in secondary and college science classrooms: A working model of helping the learning to learn.* Routledge.

Paul, G. D. (2015). Justice perceptions and practices of restorative justice facilitators and the public. *Contemporary Justice Review, 18*(3), 274–295. https://doi.org/10.1080/10282580.2015.1057678

Polytechnics Canada. (2023, July). *Polytechnic education.* https://polytechnicscanada.ca

Restorative Practice Consortium. (2017). Restorative practice resource project: Tools and successful practices for restorative schools supporting student achievement and well-being. *Restorative Practice Consortium.* https://docs.wixstatic.com/ugd/38e965_99ea6dae8cfc4db5bc42814c3c458dcc.pdf

Schneider, B. (2023). *ACIFA position on academic integrity* (Position paper). Alberta Colleges and Institutes Faculties Association. https://www.acifa.ca/_files/ugd/16fcf0_c98c6fb9f9b-c4e4880a9f875d754eb4e.pdf

Sopcak, P., & Hood, K. (2022). Building a culture of restorative practice and restorative responses to academic misconduct. In S. E. Eaton & J. Christenson Hughes (Eds.), *Academic integrity in Canada: Ethics and integrity in educational contexts* (Vol. 1, pp. 553–571). Springer. https://doi.org/10.1007/978-3-030-83255-1_29

Sternberg, R. (2013, June 17). *Giving employers what they don't really want.* Chronicle of Higher Education. http://chronicle.com/article/Giving-Employers-WhatThey/139877/?cid=ja

The Conference Board of Canada. (2014). *Innovation skills profile 2.0.* https://edata.conference-board.ca/docs/default-source/public-pdfs/InnovationSkillsProfile.pdf

Vaandering, D., & Reimer, K. (2019). Listening deeply to public perceptions of restorative justice: What can researchers and practitioners learn. *International Journal of Restorative Justice, 2*(2), 186–208.

Valandra, E. C. (2020). Undoing the first harm: Settlers in restorative justice. In E. C. Valandra (Ed.), *Colorizing restorative justice: Voicing our realities* (pp. 325–370). Living Justice Press.

Wachtel, T. (1999, February 16–18). Restorative justice in everyday life: Beyond the formal ritual (Paper presentation). Reshaping Australian Institutions Conference: Restorative Justice and Civil Society, The Australian National University, Canberra, Australia. https://www.iirp.edu/eforum-archive/4221-restorative-justice-in-everyday-life-beyond-the-formal-ritual

Zehr, H. (2003). *The little book of restorative justice*. Good Books.

Alana Fukushima (she/her) leads the Office of Student Conduct & Accountability at NAIT Polytechnic, where she oversees academic and non-academic student conduct with restorative pathway opportunities, as well as Chair's NAIT's Behavioural Intervention Team. Alana's interests lie in the intersection of people, behaviour, and structures with a focus on intentionality, community, and connection. She has written governance documents and developed programs addressing academic integrity, academic misconduct, early intervention, student mental health, crisis management, and student rights and responsibilities. Alana holds a Bachelor of Arts, double major in Psychology and Art and Design, from the University of Alberta.

Linden Couteret (they/them) is a student conduct administrator with a background in teaching, student learning success, and disability services. They have a Master's degree in education focused on developing student metacognition, a certificate in restorative justice, and a Bachelor's degree in secondary education.

Part II
Impact of Structure and Organization of Vocational and Polytechnic Education on Academic Integrity

Chapter 7
Student Stress in Community Colleges: Creating Healthy, Ethical Spaces for Learning

Corrine D. Ferguson and Margaret A. Toye

Abstract Student stress and mental health continue to be a concern for higher education institutions following the height of the COVID-19 pandemic and resulting campus closures and emergency remote teaching. Research has shown that elevated stress levels negatively impact student mental health and academic performance. Community colleges providing 8-to-24-month certificate and diploma programs, accessible through a variety of delivery modalities, attract a diverse student body with complex stress experiences. The stressors students experience in their everyday lives can also be critical motivators for engaging in academic misconduct.

Reducing stress to support mental health and academic integrity requires critically assessing the fabric of our organizations. In this chapter, we argue that the structure of community colleges may heighten student and instructor stress and pose significant barriers to creating supportive academic integrity communities. We propose that locating academic misconduct within the theoretical model of the stress process shifts attention from stress as an individual problem to how student stress can be understood in the context of the structural environment. The chapter concludes with recommendations for how community colleges may address stress and mental health using the 4M framework and apply principles of Universal Design for Learning (UDL) with consideration of equity, diversity, and inclusion (EDI) toward creating healthy, supportive, and ethical spaces for learning.

Keywords Stress · Mental health · Academic integrity community · Equity · Social support · Inclusion · UDL principles · Ethical spaces

---

C. D. Ferguson (✉) · M. A. Toye
Bow Valley College, Calgary, AB, Canada
e-mail: cferguson@bowvalleycollege.ca; mtoye@bowvalleycollege.ca

L. Vogt, B. M. Stoesz (eds.), *Academic Integrity in Vocational and Polytechnic Education*, Ethics and Integrity in Educational Contexts 8,
https://doi.org/10.1007/978-3-031-90091-4_7

Positionality and Self-Reflection

The authors fully recognize their privileged position as faculty and associate dean in a mid-size Canadian community college located at a downtown campus in a large city. Our college provides learning opportunities for students wishing to complete 8-to-24-month programs toward applied careers as well as English language learning and adult basic and high school education. We are both white and have earned higher education degrees following a Western perspective with English as our first language. We acknowledge that our positions shape the way we see the world and the way we view student work. In our roles, we provide mentorship for academic integrity and abide by and enforce the College's academic integrity policy. We did so with impunity, that is, until we started thinking about the conversations with students who were experiencing many stressful life events that seemed to act as barriers to academic integrity. It was through an exploration of student stressors that we had what we call three "awakenings" or epiphanies about stress and its relationship to integrity in applied learning settings: 1) we need only look at the ways we are trying to help students with stress to understand how we think about stress; 2) the way we think about stress is singular and positioned, and 3) turning our eyes inward and engaging in self-reflection in pursuit of change is the way forward.

In this chapter, we challenge how institutions address stress and argue that the way we organize community colleges manifests stress and produces barriers to creating academic integrity communities that support learners. This results in formidable conditions for students to complete their work with integrity. The chapter concludes with recommendations for how community colleges may address stress and mental health using the 4M framework and considers how equity, diversity, and inclusion (EDI) and Universal Design for Learning (UDL) principles can be applied toward creating healthy, supportive, and ethical spaces for learning. We begin by providing background information about student stress and the link to mental health and academic integrity.

Background

Students in higher education experience many stressors that make them vulnerable to mental health issues (Linden & Stuart, 2020) and compromised academic performance (Duffy et al., 2020). Results from the 2019 National College Health Assessment (NCHA II) state that over 45% of Canadian post-secondary students reported their level of stress as "more than average," and over 15% defined their stress levels as "tremendous" (American College Health Association, 2019). During the pandemic, these proportions rose to 47% and 36.7%, respectively (American College Health Association, 2022). Linden et al. (2021) found an upward trend (2013, 2016, 2019) in self-reported psychological distress among students and levels appeared to rise again in 2022 (American College Health Association, 2022).

Over one-half (51.5%) of students in that same year reported that stress negatively affected their academic performance (American College Health Association, 2022). Post-pandemic stress continues to be a concern for Canadian students in the face of an affordable housing crisis, food insecurity, tuition fee increases, and policy changes for international students.

In addition to mental health and academic performance, research has also shown a link between student stress and academic misconduct (McCabe, 2016; Newton, 2018). As Eaton et al. (2023a) suggest, stress and mental health concerns may arise as part of the academic integrity offence process. Pitt et al. (2021) documented student experiences during the process of contract cheating allegations, where it was described as a period of "mess, stress, and trauma" (p. 663). A study of stress at one community college revealed that students experienced a variety of stressors both in and outside the college context and that engagement in violation behaviour was significantly associated with higher stress levels (Ferguson et al., 2023). As stress is seen as a risk to mental health and academic integrity, institutions must strive to understand its manifestations in pursuit of evidence-based ways to support learners. We shift our attention now to explaining the link between stress and academic integrity behaviour by turning to theory.

Understanding the Link

Many studies have collected data to describe the frequency, nature, and type of academic integrity breach, but this information alone does not tell us all we need to know to address it in our institutions. Theory in academic integrity research is essential as "it not only *explains* observations, but it also allows *predictions* to be made about what is expected to occur" (Curtis & Clare, 2024, p.1652). In short, theory is a way for us to unpack the 'why' and 'how' of integrity violation behaviour so that we may anticipate its occurrence and work toward prevention. Theories that have addressed the relationship between stress and academic integrity largely come from psychology (e.g., theory of planned behaviour, Tindall et al., 2021) and criminology (e.g., control theory, Awdry & Groves, 2023). Their contributions toward understanding violation behaviour at the individual level have been significant.

Individual-level (micro) explanations of stress have led institutions to offer a range of supports to help students cope with stress (e.g., time management workshops; Rith-Najarian et al., 2019). While supports addressing individual skills are important, we suggest this response is singular and positioned. First, it rests on the assumption that stress is solely an individual problem, and therefore, academic integrity issues that arise because of stress happen because students do not access or master the skills that resources provided by the institution are meant to develop (e.g., failure to cope, lack of time management skills). Second, it assumes that all students have equal access to the resources or experience the resources in the same way.

Sartre (1969) described human experience in terms of a mixing of coffee and cream; once the coffee and cream combine, they become a new entity and cannot separate (as cited in Thompson et al., 2016). This analogy is useful to understand and explain individual experiences with stress in society. That is, personal factors and contextual factors become one as they are both part of the same social reality. Therefore, looking at stress and academic integrity solely as a personal issue fails to consider the importance of its social context. This perspective is what sets sociology apart from other disciplines and can change the way we think and respond to student stress in relation to academic integrity.

The stress process model (Fig. 7.1), a sociological theory, allows us to see the social context of stress and mental health. The theory, first conceptualized by Pearlin et al. (1981), observed that mental health concerns occur because of stress. The occurrence and effect of stress is structured by social and economic conditions. For example, students experiencing racism, sexism, colonization, and economic marginalization may experience greater exposure (e.g., Turner et al., 1995) and vulnerability to stressful life events (e.g., Kessler & McLeod, 1984). The very resources that could mitigate stress, such as social support, effective coping strategies, self-esteem, and mastery, tend to be viable among the more advantaged in society (Pearlin & Bierman, 2013). Recent stress process research has argued for the inclusion of ableism, disablism, stigma, and neurodiversity as stressors, which makes good sense in the context of higher education (Wolbring & Escobedo, 2023).

From the perspective of the stress process model then, stress is seen as the result of participation in social relationships, social institutions, and location within structures of advantage and disadvantage that lead to mental health concerns (Pearlin, 1989). Academic misconduct is perhaps a strategy used by students to cope with stress. Community college students are typically employed while completing their programs, are older and have familial responsibilities which produce complex stress experiences (Ferguson et al., 2023). Many learners will come into colleges differentially exposed to and with varying sensitivity to stress. *We must ask ourselves: does the way we structure our programs and institutions doubly disadvantage our students?* What follows is an examination of the ways community colleges structure

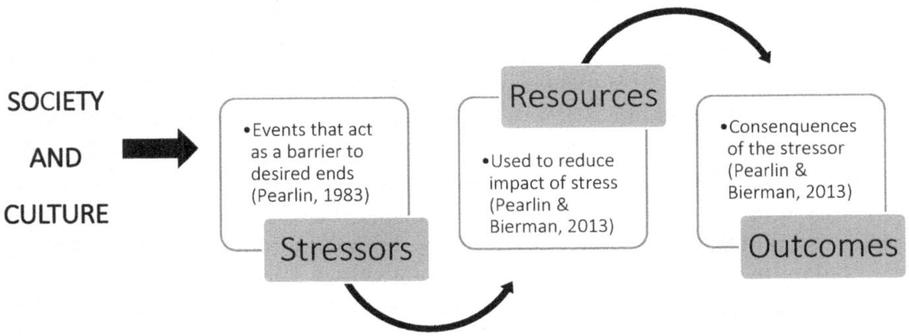

Fig. 7.1 Components of the stress process model

their learning environments that may produce the conditions for or exacerbate stress and make it difficult to build an academic integrity community to support learners.

A Closer Look

We turn now to exploring characteristics of community colleges that may contribute to stress for students and instructors, potentially leading to challenging learning and working experiences as well as difficulties in creating an academic integrity community. Canadian community colleges serve many functions. They offer English language training, adult basic and high school education, and professional training for employment in specific sectors, such as nursing, human services, information technology services, administrative services, construction trades, engineering technologies, and business careers. Professional training is often offered through 8–24-month programs. We thus consider community colleges' access mandate, instructor experience and employment status, multiple delivery modes, the structure of college programs, and decentralized academic integrity support.

Access Institutions

Community colleges in Canada have an access mandate (Dennison & Gallagher, 1986). Many colleges admit students based on the timing of their application (i.e., first applied/qualified first accepted) rather than through a competitive admission process, and programs set their admission requirements to the minimum academic and language proficiency standards needed to be successful in the program and in the workplace. This ensures that education is accessible to a diverse student population.

One of the challenges that results from this is a wide range of communication and writing skills demonstrated by students in a single course or program. Their ease, confidence, and skills in writing in the language of instruction vary, as do their experiences with citing and referencing. Ferguson et al.'s (2022) study of students in post-secondary programs at a community college showed less than half of survey respondents reporting that they were "very confident" about how to cite and reference sources in a written assignment (47.8%) and about paraphrasing or summarizing an author's words (49.5%). This results in instructional and student support resources being heavily focused on working with students to achieve a baseline of writing skills, leaving little capacity remaining for building a culture of academic integrity.

A related challenge results from students coming from countries all over the world—having immigrated to Canada or in the country as international students—and thus from a range of academic structures and systems. As Bretag (2015) notes, there are "a variety of interpretations of [academic integrity] as a concept" (p. 3) around the world, and newcomers' understanding of academic integrity does not always align with that of their Canadian post-secondary institution. Indigenous students' understandings may also be different; Lindstrom (2022) highlights that the concept of integrity, in an Indigenous paradigm, is holistic and part of all areas of life, whereas in a Western paradigm, integrity can be fragmented into many pieces, including academic integrity (p. 126). Steeves (2024) contends that marginalized students can feel that their ways of knowing are less valued than Western ways (p. 1).

Academic integrity is also addressed differently in the context of English language training, high school and adult basic education courses, and professional training, potentially making it difficult to create an institution-wide academic integrity community. Challenges also arise as students progress through English language or basic education into high school and post-secondary courses; how can the institution prepare students for its post-secondary expectations of academic integrity throughout a student's educational journey? There are complications, such as provincial high school curriculum using one citation system (e.g., MLA) and the post-secondary curriculum at the same institution using another (e.g., APA). All of this may produce stress as students navigate higher expectations for integrity and a new set of guidelines.

Instructor Experience & Employment Status

Instructors in career-related programming at community colleges are largely recruited from industry and hired for their knowledge and experience in their sector rather than for their teaching and assessing practice. Instructors may thus have varied familiarity with and expectations regarding academic integrity, leading to inconsistent standards within one program and leaving students stressed about adhering to the possibly conflicting requirements of their different instructors.

Instructors also experience stress regarding expectations around academic integrity. Two studies of faculty at individual Alberta community colleges illustrate this: Openo and Robinson (2021) describe instructors' sense of frustration when dealing with instances of academic integrity violations as well as a sense of alienation from other instructors and the institution as a result of the lack of consistency regarding academic integrity (pp. 28–29), and Hamilton and Wolsky (2022) show that fear (of e.g., damaging relationships, confrontation, retaliation by students, reputational damage, and job loss) leads to some instructors choosing not to report academic integrity violations (p. 476).

These stressful experiences and resulting emotions may disincentivize instructors from building their knowledge of academic integrity and being part of a campus-wide academic integrity community. Yet such a community is needed, as contended by the Alberta Colleges and Institutes Faculties Associations' (2023) position paper on academic integrity, which states that colleges are required to promote academic integrity culture not only for quality assurance but also "because it engenders trust in institutions and ensures students are graduating with credentials that are valued by employers and backed by an institution's reputation" (p. 2).

As Schneider (2025) addresses, another barrier to fostering academic integrity communities is that many instructors are hired on contracts and paid per course or per teaching hour and may not be paid for the time needed to engage with academic integrity. Although contract instructors aiming for full-time employment may be motivated to explore academic integrity violations and participate in academic integrity communities to impress hiring managers, others may not feel the incentive to be part of such. Crossman (2019) contends, "arguably the extra time commitments to deal with academic misconduct –as opposed to choosing to overlook it – are beyond the scope of the terms of employment. This is surely a deterrent to developing a cultural [sic] of integrity" (p. 36).

Delivery Modality

As access institutions, community colleges offer students many options for how they may complete their courses by providing a variety of delivery modalities. Many students choose the online option so they may complete their programs while working to support their education and often, their families. Sublett (2022) refers to online learning in community colleges as an "access-equity dualism" (p. 27). Although offering online courses opens the door to higher education, it may, however, create additional challenges once inside. Asynchronous online courses allow students to choose when and where they complete their academic work. This creates flexibility for students who, for varying reasons, must remain at home (e.g., primary caregivers or those recovering from health issues) and for those who are geographically distant. Despite the benefits, students may experience barriers to online learning, including access to technology and high-speed internet, challenging personal environments, and lack of community. Students also face equity issues in relation to communication when they have fewer opportunities to connect with instructors and reduced access to critical resources such as supports to promote mental and physical well-being. All these barriers may pose a risk to student health and well-being (Brown et al., 2023; Sublett, 2022). In addition to access-equity issues, methods to monitor the integrity of online assessment may also produce stressful conditions. Popularized during the period of remote learning in response to the COVID-19 pandemic, many institutions use e-proctoring software to monitor academic integrity in online assessments. More recent studies on the use of remote proctoring have uncovered inequitable assessment environments and stress-related mental health concerns (Marano et al., 2024; Pokorny et al., 2023). Vulnerable learners then may be doubly disadvantaged as online learning, although easily accessed, may produce conditions that exacerbate stress and compromise the support needed to help mitigate stress and promote academic integrity.

Program Structure

Community colleges provide professional training for graduates to enter the workforce and do so in a relatively short amount of time (i.e., 8–24 months). Many students appreciate the shorter length of a certificate or diploma program and are motivated to complete it and begin working and earning. New graduates fulfill the demand for workers in many sectors, and in the province of Alberta, the government incentivizes post-secondaries to prepare learners for those sectors through a performance metric in the Investment Management Agreement. A certain amount of base operating grant funding for each post-secondary is based on the "Proportion of recent graduates in employment two years after graduation whose current main job is very or somewhat related to the general skills and abilities acquired during their program" (Government of Alberta, 2022, p. 2).

Certificate and diploma programs provide learners with a comprehensive foundation in the sector or specific role of their future employment. As such, programs cover a breadth of curriculum. Many community college programs are accredited by industry bodies, which review programs to ensure that students are gaining all the competencies required in the sector. Programs are thus often structured with a higher course load in a term than students would experience at a university in the same province. It is not unusual for a college program to include six 3-credit courses in a term, whereas a full course load at university is considered five 3-credit courses. College programs are also apt to include both work-integrated learning experiences and traditional classes in the same term; a student may be spending, for example, two days a week at a practicum site, attending a regular practicum seminar, and completing assignments, while additionally completing four 3-credit classes.

This high course load is a stressor that characterizes the college environment. Ferguson et al.'s (2023) study identified "increased academic workload" (p. 697) as the second most common stressor students experience. "Fear of not graduating" (p. 697) was the sixth most common stressor, which reflects the pressure students feel to complete the course requirements and get working. The strategy of taking fewer courses at a time to mitigate workload stress leads to students taking longer to enter the workforce, which in itself could be a stressor as it could result in financial hardship. Also, students with funding or who have study permits are often required to take a full course load in order to maintain full-time student status. The option to reduce their course load is not available to them, which can certainly contribute to their experience of stress.

Decentralized Academic Integrity Support

Academic integrity work in higher education, despite being considered fundamental, is often spread across the institution and completed in the "third space" (Ahuna et al., 2023). Third space labour refers to academic integrity work as a "side gig" (Ahuna et al., 2023, p. 78), distributed widely across multiple roles at the institution in transitory places and spaces. In their analysis of academic integrity position postings in Canada, the United States, and Australia, Vogt and Eaton (2022) note ten different departments across post-secondary education where positions are located, including, for example, student services, the library, and the teaching and learning centre. Academic Integrity Offices are not found in many Canadian educational institutions, and how they function to support integrity is also variable (Vogt & Eaton, 2022). While some colleges and polytechnics have a centralized academic integrity office to support students and/or instructors, others do not, leaving the primary responsibility for enforcing, monitoring, and reporting in the hands of instructors and program chairs (at the author's institution, these are instructors with course relief who oversee and support teaching in each program area). Academic integrity is often not the employee's sole role, and they will hold many other responsibilities in addition to supporting academic integrity (Ahuna et al., 2023). Additionally, they may not have received specialized training to implement institutional academic integrity policy or conduct investigations (Eaton et al., 2023b). Assigning primary detection, investigation, and reporting to instructors in colleges produces instructor stress in the form of increased workload and added emotional burden associated with academic integrity work (Openo & Robinson, 2021; Schneider, 2025). Instructors must also navigate a relationship with the student after administering academic integrity policy procedures, forming a critical stress point that may interrupt the continuity and stability of the mentor relationship and threaten student and instructor mental well-being.

Academic integrity education/training may also be decentralized with different departments supporting various community members (Vogt & Eaton, 2022). This decentralized model may be insufficient to manage policy and result in inconsistent applications within educational institutions. With no central direction there may be considerable variability in the interpretation, application, and approaches to academic integrity across instructors, courses, and programs. This inconsistency produces student stress as it makes it difficult for learners to anticipate the course-specific expectations to complete their work with integrity, and to predict the possible responses to breaches. In Eaton et al.'s (2023b) study, faculty identify having dedicated offices and staff to promote academic integrity and provide education as a key facilitator in promoting integrity and preventing academic misconduct. A centralized office can also function as a primary source of support through which instructor and, to some extent, student stress, may be mitigated. See Lucien-Bay et al. (2025) for applied insights into the role of student conduct officers who support faculty and chairs through centralized offices.

Recommendations

The preceding section focused on how the structure and organization of a community college may create or intensify stress and challenge community building, putting academic integrity and mental health at risk. In this section, we propose recommendations for how community colleges may consider the systemic impacts on stress and mental health and move toward creating healthy, supportive, and ethical spaces for learning. To do this, we draw upon the 4M framework and consider equity, diversity, and inclusion (EDI), and Universal Design for Learning (UDL) principles.

Key figures in academic integrity research have long called for using a holistic approach to fostering integrity at our institutions (e.g., Bertram Gallant, 2008; Bretag et al., 2014; Macdonald & Carroll, 2006). According to Bretag et al. (2014), a holistic approach would involve:

> promoting integrity in every aspect of the academic enterprise: including university mission statements and marketing, through admissions processes, to nuanced and carefully articulated policy. It must include assessment practices and curriculum design, information provided during orientation, and frequent and visual reminders on campus. There must be embedded and targeted support in courses and at every level for students, professional development for staff, and research training (Bretag et al., 2014, p. 1153).

Influenced by systems theory, Kenny and Eaton (2022) applied the 4M framework from Scholarship of Teaching and Learning (SoTL) discourse to academic integrity. The premise of the framework is that academic integrity is the responsibility of multiple stakeholders, including students, educators, and administrators. The framework consists of four interrelated levels: micro (individual behaviours and decisions), meso (departmental, faculty, and committee activities creating culture and practices), macro (broader institutional structures and policies), and mega (societal and cultural factors) (Kenny & Eaton, 2022). All levels play a critical role in building a culture of integrity. By extension, we argue that all levels of an institution are also responsible for the mental well-being of its members. Consideration of EDI, including implementation of UDL principles in pursuit of academic integrity and mental well-being, must, therefore, be undertaken at all levels.

EDI is finding its place in many applied learning settings, being widely employed to adjust policy, student supports, and accommodations. At the authors' own institution, understandings of EDI are evolving, and we use a current version here to build a collective understanding. Equity refers to the acknowledgement and removal of structural and procedural barriers, while diversity is "the community composition that employees and learners, especially equity-deserving populations, view as representative, trustworthy, and accountable" (Bow Valley College, 2024). Inclusion is realized through actions that promote equity and meet the needs of all in the community (Bow Valley College, 2024). To this end, UDL (CAST, 2018) principles focus on making learning inclusive or more accessible so that all students may participate in meaningful learning opportunities. UDL has gained traction in teaching and learning (e.g., Brown et al., 2023; Coleman, 2022); principles are being used in assessment (e.g., Estefan et al., 2023) and course design (e.g., Kumar & Wideman, 2014), and to promote academic integrity (e.g., Norris et al., 2024).

Table 7.1 provides examples of how community colleges can respond to the characteristics contributing to student and instructor stress using EDI and UDL principles. These ideas are our own or drawn from relevant research. The examples of innovative solutions are gleaned from strategies described by other authors in this

Table 7.1 Recommendations to Address Stress and Mental Health

Community College Characteristic	EDI Principles	UDL Principles
Access Institution		
• Wide range of writing skills • Varying confidence in writing in the language of instruction • Variety of interpretations of academic integrity • Conflicting expectations across learning pathways	**Decolonize and Indigenize**[a] academic integrity policy, procedure, resources, and teaching practices. (Macro; Meso) Consider multiple understandings of integrity and multiple ways to demonstrate integrity while promoting relationship building toward a supportive community (Lindstrom, 2022; Poitras Pratt & Gladue, 2022). *Example:* A collaboration between Lorisia MacLeod and NorQuest Indigenous Student Centre developed citation templates for Indigenous Elders and Knowledge Keepers (MacLeod, 2021).	**Graduated and focused academic integrity support** to aid in the transition both from high school to college and from English language learning (ELL) programs to college-level programs. (Macro) *Example:* The Academic Readiness Program (SAIT) provides educational support to international students through academic integrity modules that are designed with an intercultural perspective (El Chaar et al., 2024). **Flexibility in policy and procedure** to allow for academic integrity learning for students in ELL programming. (Macro) **Scaffolding** to promote academic integrity learning throughout a course. (Meso) **Provide feedback** that encourages perseverance, development of confidence around citing and referencing, and use of specific supports. (Meso) **Celebrate diverse understandings**. Promote academic integrity understanding across cultures. (Parnther, 2022) (Meso) *Example:* Encourage students to explore how Western and non-Western perspectives of academic integrity connect (Ferguson et al., 2024).
Instructor Experience & Employment Status		
• Varying knowledge of academic integrity • Lack of incentive to build knowledge	**Bridge the divide**. Include academic integrity training when onboarding for both contract and continuous instructors. Training should be offered as a paid endeavour delivered in multiple formats so casual instructors working in other paid employment with time constraints can access the training. (Macro) Further program-specific training would provide necessary context to academic integrity issues. (Meso) *Example:* Scurr (2025) describes an academic integrity micro-credential faculty can take as part of their required teacher certification.	**Develop a community of practice** to support instructors/staff in developing their knowledge of academic integrity and shape the discourse around academic integrity at the institution (Kenny & Eaton, 2022; Verhoef et al., 2022). (Meso) *Example:* Bow Valley College has created the Academic Integrity Learning Network (AILN) accessed by instructors, program chairs, and staff in MS Teams. It provides relevant research, teaching and learning practices, and professional development opportunities, creating a hub for questions and dialogue. **Reward engagement.** Participation in the academic integrity community could also be incentivized. *Example:* Providing course release in return for serving as an academic integrity representative for a program area.

(continued)

Table 7.1 (continued)

Community College Characteristic	EDI Principles	UDL Principles
Delivery Modality		
• Access to device/ internet • Personal environment • Loss of community • Loss of access to critical resources • E-proctoring surveillance	**Support digital, space, and resource equity.** To address technology barriers, offer loaner devices, flexible access to computer labs, and innovative solutions to accessing costly software. (Macro) *Example:* Bow Valley College created a data centre to allow students access to "elite" processing power. Students are provided with a high-powered virtual desktop they can access with any device (i.e., smartphones, tablets, and computers) at no extra cost. Equity of communication and access to critical resources to support mental health. Ensure students across all delivery modalities have a variety of ways to connect with other students and with instructors. Resources to support mental health should be culturally specific, when possible, free and available to access in multiple ways and at different times for students who may be completing courses outside of business hours (Sublett, 2022). (Meso) Equity in access to academic integrity resources, knowledge sharing, and training with respect to policy and procedure. (Macro; Meso) *Example*: Assiniboine Community College has created a Library Guide for students that includes a link to the academic integrity policy, easy-to-understand descriptions of behaviours that violate academic integrity, as well as supports to help students understand and master the skills to complete their work with integrity (https://assiniboine.libguides.com/ai-students).	**Flexibility is key.** Space out assignments to avoid crunch times. Avoid stay-up-late deadlines. Provide flexibility in resubmission policies (e.g., rewriting, ungrading). Offer optional late bank days (e.g., five days) so students can choose to submit assessments after the deadline with no penalty and no questions asked. Allow for multiple formats to express learning. (Meso) Instructor training in suicide and mental health support (Coleman, 2022). **Create a class community.** Build a class community to support open inquiry and support. Schedule connect times or office hours. Implement an academic integrity buddy system where students can help one another with citing and referencing. (Meso) **Make a space.** Design an academic integrity hub in your course that provides learners with information displayed in a variety of formats (i.e., visual, audio). Allow anonymous posts for students to ask questions to avoid negative stigma. (Meso) **Consider accessibility**. Provide accessible (i.e., 24-hour, free) mental health supports for online learners who may not be able to get to campus or who live in geographically more remote areas. (Meso) **Close the camera and open the book.** Offer alternative expressions of learning to standard multiple choice or provide open-book exams. (Meso) *Example:* Conestoga College has a helpful resource on best practices for open-book testing (Stoutenburg, 2023).
Program Structure		
• Course load • Pressure to complete	**Level the field.** Design programs and courses with time in mind, as students who are economically marginalized feel the pressure to complete their programs more acutely. *Example*: Offer flexible completion times for programs. (Meso) and consider a maximum number of courses per term. (Meso)	**Implement strategies to lower the pressure**. Focus on strategies to optimize flexibility and choice, such as dropped grading (i.e., best x out of y) and late bank days with no questions asked. Offer a variety of ways to express learning. (Meso) **To group or not to group?** That is a question the students can answer. Provide them with a choice to opt out of group projects when course outcomes allow. (Meso)

(continued)

Table 7.1 (continued)

Community College Characteristic	EDI Principles	UDL Principles
Decentralized Support		
• Inconsistent application of policy • Instructor workload • Emotional burden • Instructor-student relationship	**Create an independent academic integrity office.** As an impartial third party, an academic integrity office manages the investigation and reporting which may create greater consistency across the institution, promote an institutional culture of integrity and reduce instructor workload and emotional burden (Eaton et al., 2023b). (Macro) **Anchor academic integrity support.** A centralized physical and virtual space for academic integrity support to address equity in access to academic integrity resources, knowledge sharing, and training. (Macro) *Example*: The University of Waterloo's Office of Academic Integrity has both a physical and virtual presence. Their website is a learning hub for instructors, students, and staff. The office provides comprehensive resources and offers educational workshops. They send a centralized and consistent message about academic integrity to the three groups, which reinforces the idea that it takes a community to build integrity. (Meso; Macro) (https://uwaterloo.ca/academic-integrity/).	**Make academic integrity friends.** Engage in collaboration with other supporters of academic integrity within your institution (McKenzie, 2024) (Meso; Macro; Mega) **Draw attention to academic integrity workload issues** in employment agreements (Schneider, 2025). (Macro) **Encourage instructor-student partnerships** in academic integrity endeavours at your institution and beyond. Building stronger, more equitable relationships between students and instructors can mend relationships (Micro), improve academic integrity (Bertram Gallant, 2017), and promote an institutional culture of integrity (Lancaster, 2022). (Macro). *Example:* Include students in academic integrity working groups, the research ethics board, and student-led academic integrity workshops. Apply students as partners methodology in teaching and research activities. Offer opportunities for students to volunteer or work as peer tutors (Ferguson et al., 2024). (Meso) Present a poster together at an academic integrity conference (Aburto & Ferguson, 2024). (Mega)

[a]There are many different definitions of these terms. We embrace the definitions offered by Eaton (2024) that draws on important ideas and frameworks from Tuck and Yang (2012) and Gladue (2020a, 2020b). Decolonization "includes recognizing, challenging and then breaking down systems of oppression rooted in colonial practices" (Gladue, 2020a, 2020b as cited in Eaton, 2024, p. 3). Whereas Indigenization refers to work done by Indigenous peoples to introduce Indigenous ways of knowing to colonial systems (Gladue, 2020b as cited in Eaton, 2024). For example, challenging colonialist ideals of unbiased scholarly research may be considered a decolonization practice in teaching and learning. While including Indigenous principles (e.g., relationality, reciprocity, respect) in institutional definitions of academic integrity by Indigenous peoples constitutes Indigenization (Gladue & Poitras Pratt, 2024)

book or from learning institutions across Canada and elsewhere. Supports to mitigate stress are usually applied at the micro level, but we recommend changes at the meso, macro, and mega levels as they have great potential to mitigate student and instructor stress and build academic integrity communities. We would be remiss to overlook the challenges academic integrity supporters face in advocating for change at their institutions. Seeland et al. (2024) underscore the difficulty of producing change, particularly at the meso level, due to resource and time constraints and suggest engaging in informal collaboration and forming networks to act as a conduit between the meso and macro levels to leverage change.

Conclusions

In this chapter, we underscore the importance of addressing student and instructor stress to promote mental well-being and academic integrity. We argue that the very way we structure our learning environments produces stress and this may create difficult conditions for students to complete their work with integrity. In the absence of inclusive and accessible social support, these conditions are ripe for academic misconduct. The authors share a selection of actions to address stress and promote support within college settings, but this is by no means an exhaustive list. We learn from all of you, and we hope you have learned from us.

In the spirit of "What If," a song that muses about how our world would change in response to radical shifts in thinking (Williams, 2007):

> What if … we shift the way we think about stress?
>
> What if … we turn our gaze inward and critically examine how the structure of our learning environments disadvantages our students?
>
> What if … we change the environments in a way that addresses stress and promotes mental well-being?
>
> We would create healthy, supportive, and ethical spaces for learning.

References

Aburto, S., & Ferguson, C. D. (2024, March 9). *Art, artificial intelligence, and academic integrity: Lessons learned in sociology class* (Poster presentation). International Center for Academic Integrity (ICAI) Annual Conference 2024, Calgary, Alberta, Canada.

Ahuna, K. H., Frankovitch, L., & Murphy, G. (2023). Claiming space for honest work: Academic integrity as third space labor. *Workplace: A Journal for Academic Labor, 34*, 75–87. https://doi.org/10.14288/workplace.v34i

Alberta Colleges and Institutes Faculties Association. (2023, March 17). *ACIFA's position on academic integrity*. https://www.acifa.ca/_files/ugd/16fcf0_c98c6fb9f9bc4e4880a9f875d754eb4e.pdf

American College Health Association. (2019). *American College Health Association-National College Health Assessment II: Canadian Consortium Executive, Summary Spring 2019*. American College Health Association. https://www.acha.org/documents/ncha/NCHA-II_SPRING_2019_US_REFERENCE_GROUP_EXECUTIVE_SUMMARY.pdf

American College Health Association. (2022). *American College Health Association-National College Health Assessment III: Canadian reference group executive summary spring 2022*. American College Health Association. https://www.acha.org/documents/ncha/NCHA-III_SPRING_2022_REFERENCE_GROUP_EXECUTIVE_SUMMARY.pdf

Awdry, R., & Groves, A. (2023). Why they do and why they don't: A combined criminological approach to understanding assignment outsourcing in higher education. *International Journal for Educational Integrity, 9*(7). https://doi.org/10.1007/s40979-023-00126-3

Bertram Gallant, T. (2008). Academic integrity in the twenty-first century: A teaching and learning imperative. *ASHE Higher Education Report, 33*(5), 1–143. https://doi.org/10.1002/aehe.3305

Bertram Gallant, T. (2017). Academic integrity as a teaching and learning issue: From theory to practice. *Theory into Practice, 56*(2), 88–94. https://doi.org/10.1080/00405841.2017.1308173

Bow Valley College. (2024, March 5). What does equity, diversity, and inclusion mean to Bow Valley College? *Equity, Diversity, Inclusion*. https://bowvalleycollege.ca/student-resources/student-life/equity-diversity-and-inclusion

Bretag, T. (2015). Defining academic integrity – international perspectives: Introduction. In T. Bretag (Ed.), *Handbook of academic integrity*. Springer. https://doi.org/10.1007/978-981-287-098-8_76

Bretag, T., Mahmud, S., Wallace, M., Walker, R., McGowan, U., East, J., Green, M., Partridge, L., & James, C. (2014). 'Teach us how to do it properly!' An Australian academic integrity student survey. *Studies in Higher Education, 39*(7), 1150–1169. https://doi.org/10.1080/03075079.2013.777406

Brown, M., Hoon, A. E., Edwards, M., Shabu, S., Okoronkwo, I., & Newton, P. M. (2023). A pragmatic evaluation of university student experience of remote digital learning during the COVID-19 pandemic, focusing on lessons learned for future practice. *PLoS One, 18*(5), e0283742. https://doi.org/10.1371/journal.pone.0283742

CAST (2018). *Universal design for learning guidelines version 2.2*. http://udlguidelines.cast.org

Coleman, M. E. (2022). Mental health in the college classroom: Best practices for instructors. *Teaching Sociology, 50*(2), 168–182. https://doi.org/10.1177/0092055X221080433

Crossman, K. (2019). Is this in my contract? How part-time contract faculty face barriers to reporting academic integrity breaches. *Canadian Perspectives on Academic Integrity, 2*(1), 32–39. https://doi.org/10.11575/cpai.v2i1.68934

Curtis, G., & Clare, J. (2024). Academic integrity scholarship: The importance of theory. In S. E. Eaton (Ed.), *Second handbook of academic integrity* (pp. 1651–1669). Springer. https://doi.org/10.1007/978-3-031-54144-5_164

Dennison, J. D., & Gallagher, P. (1986). *Canada's community colleges: A critical analysis*. UBC Press.

Duffy, A., Keown-Stoneman, C., Goodday, S., Horrocks, J., Lowe, M., King, N., et al. (2020). Predictors of mental health and academic outcomes in first-year university students: Identifying prevention and early-intervention targets. *BJPsych Open, 6*(3), 1–8. https://doi.org/10.1192/bjo.2020.24

Eaton, S. E. (2024). Decolonizing academic integrity: Knowledge caretaking as ethical practice. *Assessment & Evaluation in Higher Education, 49*, 1–16. https://doi.org/10.1080/02602938.2024.2312918

Eaton, S. E., Pethrick, H., & Turner, K. L. (2023a). Academic integrity and student mental well-being: A rapid review. *Canadian Perspectives on Academic Integrity, 5*(2), 34–58. https://doi.org/10.11575/cpai.v5i2.73748

Eaton, S. E., Stoesz, B. M., Crossman, K., Garwood, K., & McKenzie, A. (2023b). Faculty perspectives of academic integrity during COVID-19: A mixed methods study of four Canadian universities. *Canadian Journal of Higher Education/Revue Canadienne D'enseignement Supérieur, 52*(3), 42–58. https://doi.org/10.47678/cjhe.vi0.189783

El Chaar, D., Radke, S., Bernard, K., & De Leon, A. (2024, March 8). *Addressing academic misconduct among international students: An intercultural approach* (Conference presentation). International Center for Academic Integrity (ICAI) Annual Conference, 2024, Calgary, Alberta, Canada.

Estefan, M., Selbin, J. C., & Macdonald, S. (2023). From inclusive to equitable pedagogy: How to design course assignments and learning activities that address structural inequalities. *Teaching Sociology, 51*(3), 262–274. https://doi.org/10.1177/0092055X231174515

Ferguson, C. D., Toye, M. A., Carver, C., Pictin, T., Eaton, S. E., & Boisvert, S. (2022). *Understanding student experiences of commercial contract cheating and other outsourcing behaviours – Research report*. Bow Valley College. https://prism.ucalgary.ca/handle/1880/115343

Ferguson, C. D., Toye, M. A., & Eaton, S. E. (2023). Contract cheating and student stress: Insights from a Canadian community college. *Journal of Academic Ethics, 21*, 685–717. https://doi.org/10.1007/s10805-023-09476-6

Ferguson, C. D., Toye, M. A., & Dyer, F. (2024, March 9). *Building community: Creating faculty and staff – student partnerships at a Canadian applied learning college* (Poster presentation).

International Center for Academic Integrity (ICAI) Annual Conference 2024, Calgary, Alberta, Canada. https://hdl.handle.net/1880/118586

Gladue, K. (2020a, November 13). *Indigenous academic integrity* [PDF]. https://taylorinstitute.ucalgary.ca/sites/default/files/Content/Resources/Academic-Integrity/21-TAY-Indigenous-Academic-Integrity.pdf

Gladue, K. (2020b, November 25*). Indigenous paradigms in practice: Relationships, story and academic integrity* (Video). Academic integrity: Urgent and emerging topics webinar series, Taylor Institute for Teaching and Learning, University of Calgary. https://youtu.be/I0ZJl3dLmKg

Gladue, K., & Poitras Pratt, Y. (2024). The integrity of good relations: Indigenous approaches to respect, relationality, and reciprocity in higher learning. In S. E. Eaton (Ed.), *Second handbook of academic integrity* (pp. 1223–1244). Springer. https://doi.org/10.1007/978-3-031-54144-5_132

Government of Alberta. (2022). *2022–2025 investment management agreement: Bow Valley College*. Government of Alberta. https://open.alberta.ca/dataset/358a9125-98a4-4a8d-bac1-d76995519ecd/resource/ef6361bb-92c8-4ad5-9a8e-d75fd28f098f/download/ae-investment-management-agreement-2022-2025-amended-and-restated-bvc.pdf

Hamilton, M., & Wolsky, K. (2022). The barriers to faculty reporting incidences of academic misconduct at community colleges. In S. E. Eaton & J. Christensen Hughes (Eds.), *Academic integrity in Canada. Ethics and integrity in educational contexts, 1* (pp. 467–484). Springer. https://doi.org/10.1007/978-3-030-83255-1_24

Kenny, N., & Eaton, S. E. (2022). Academic integrity through a SoTL Lens and 4M framework: An institutional self-study. In S. E. Eaton & J. Christensen Hughes (Eds.), *Academic integrity in Canada. Ethics and integrity in educational contexts, 1* (pp. 573–592). Springer. https://doi.org/10.1007/978-3-030-83255-1_30

Kessler, R. C., & McLeod, J. D. (1984). Sex differences in vulnerability to undesirable life events. *American Sociological Review, 49*(5), 620–631.

Kumar, K. L., & Wideman, M. (2014). Accessible by design: Applying UDL principles in a first year undergraduate course. *The Canadian Journal of Higher Education, 44*(1), 125–147. https://doi.org/10.47678/cjhe.v44i1.183704

Linden, B., & Stuart, H. (2020). Post-secondary stress and mental wellbeing: A scoping review of the academic literature. *Canadian Journal of Community Mental Health, 39*(1), 1–32. https://doi.org/10.7870/cjcmh-2020-002

Linden, B., Boyes, R., & Stuart, H. (2021). Cross-sectional trend analysis of the NCHA II survey data on Canadian post-secondary student mental health and wellbeing from 2013 to 2019. *BMC Public Health, 21*(1), 1–13. https://doi.org/10.1186/s12889-021-10622-1

Lindstrom, G. (2022). Accountability, relationality and indigenous epistemology: Advancing an indigenous perspective on academic integrity. In S. E. Eaton & J. Christensen Hughes (Eds.), *Academic integrity in Canada: An enduring and essential challenge* (pp. 125–139). Springer. https://doi.org/10.1007/978-3-030-83255-1_6

Lucien-Bay, L., Couteret, L., & Radke, S. (2025). Polytechnic pinch points for academic integrity: Reflections from student conduct officers. In L. Vogt & B. M. Stoesz (Eds.), *Academic integrity in vocational and polytechnic education*. Springer.

Macdonald, R., & Carroll, J. (2006). Plagiarism: A complex issue requiring a holistic institutional approach. *Assessment and Evaluation in Higher Education, 1*(3), 233–245. https://doi.org/10.1080/02602930500262536

MacLeod, L. (2021). More than personal communication: Templates for citing indigenous elders and knowledge keepers. *KULA: Knowledge Creation, Dissemination, and Preservation Studies, 5*(1). https://doi.org/10.18357/kula.135

Marano, E., Newton, P. M., Birch, Z., Croombs, M., Gilbert, C., & Draper, M. J. (2024). What is the student experience of remote proctoring? A pragmatic scoping review. *Higher Education Quarterly, 78*(3), e12506. https://doi.org/10.1111/hequ.12506

McCabe, D. (2016). Cheating and honor: Lessons from a long-term research project. In T. Bretag (Ed.), *Handbook of academic integrity* (pp. 187–198). Springer. https://doi.org/10.1007/978-981-287-098-8_35

McKenzie, A. (2024, March 9). *#MakeItSomeonesJob: Building an academic integrity office* (Conference presentation). International Centre for Academic Integrity (ICAI) Annual Conference, 2024, Calgary, Alberta, Canada.

Newton, P. M. (2018). How common is commercial contract cheating in higher education, and is it increasing? A systematic review. *Frontiers in Education, 3*(67). https://doi.org/10.3389/feduc.2018.00067

Norris, M. E., Karasewich, T. A., & Kenkel, H. K. (2024). Academic integrity and accommodations: The intersections of ethics and flexibility. In S. E. Eaton (Ed.), *Second handbook of academic integrity* (pp. 249–268). Springer. https://doi.org/10.1007/978-3-031-54144-5_92

Openo, J. A., & Robinson, R. (2021). The emotional labour of academic integrity: How does it feel? *Canadian Perspectives on Academic Integrity, 4*(1), 20–39. https://doi.org/10.11575/cpai.v4i1.71350

Parnther, C. (2022). International students and academic misconduct: Considering culture, community and context. *Journal of College and Character, 23*(1), 60–75. https://doi.org/10.1080/2194587X.2021.2017978

Pearlin, L. I. (1983). Role strains and personal stress. In H. B. Kaplan (Ed.), *Psychological stress: Trends in theory and research* (pp. 3–30). Academic Press.

Pearlin, L. I. (1989). The sociological study of stress. *Journal of Health and Social Behavior, 30*(3), 241–256.

Pearlin, L. I., & Bierman, A. (2013). Current issues and future directions in research into the stress process. In C. S. Aneshensel, J. C. Phelan, & A. Bierman (Eds.), *Handbook of the sociology of mental health* (2nd ed., pp. 325–340). Springer Science + Business Media. https://doi.org/10.1007/978-94-007-4276-5_16

Pearlin, L. I., Menaghan, E. G., Lieberman, M. A., & Mullan, J. T. (1981). The stress process. *Journal of Health and Social Behavior, 22*(4), 337–356. https://doi.org/10.2307/2136676

Pitt, P., Dullaghan, K., & Sutherland-Smith, W. (2021). "Mess, stress and trauma": Students' experiences of formal contract cheating processes. *Assessment & Evaluation in Higher Education, 46*(4), 659–672. https://doi.org/10.1080/02602938.2020.1787332

Poitras Pratt, Y., & Gladue, K. (2022). Re-defining academic integrity: Embracing indigenous truths. In S. E. Eaton & J. Christensen Hughes (Eds.), *Academic integrity in Canada: An enduring and essential challenge* (pp. 103–123). Springer. https://doi.org/10.1007/978-3-030-83255-1_5

Pokorny, A., Ballen, C. J., Drake, A. G., Driessen, E. P., Fagbodun, S., Gibbens, B., Henning, J. A., McCoy, S. J., Thompson, S. K., Willis, C. G., & Lane, A. K. (2023). "Out of my control": Science undergraduates report mental health concerns and inconsistent conditions when using remote proctoring software. *International Journal for Educational Integrity, 19*(22). https://doi.org/10.1007/s40979-023-00141-4

Rith-Najarian, L. R., Mesri, B., Park, A. L., Sun, M., Chavira, D. A., & Chorpita, B. F. (2019). Durability of cognitive behavioral therapy effects for youth and adolescents with anxiety, depression, or traumatic stress: A meta-analysis on long-term follow-ups. *Behavior Therapy, 50*(1), 225–240. https://doi.org/10.1016/j.beth.2018.05.006

Sartre, J. P. (1969). *Being and nothingness: An essay on phenomenological ontology*. Methuen.

Schneider, B. (2025). Advocating for fair compensation and employment equity: Reducing the burden of academic integrity and misconduct work in colleges and polytechnics. In L. Vogt & B. M. Stoesz (Eds.), *Academic integrity in vocational and polytechnic education*. Springer.

Scurr, C. (2025). Faculty education on academic integrity: A micro-credential approach. In L. Vogt & B. M. Stoesz (Eds.), *Academic integrity in vocational and polytechnic education*. Springer.

Seeland, J., Bateman, KC., Eaton, S. E., Bens, S., & Rouse, A. (2024, March 9). *Constructing academic integrity: Pillars of institutional culture* (Panel presentation). International Center for Academic Integrity Annual Conference (ICAI), 2024, Calgary, Alberta, Canada.

Steeves, M. (2024). Addressing equity in academic integrity. *Canadian Perspectives on Academic Integrity, 7*(2), 1–7. https://doi.org/10.55016/ojs/cpai.v7i2/77686

Stoutenburg, L. (2023, July 13). *Best practices for open book testing*. https://tlconestoga.ca/best-practices-for-online-open-book-testing/

Sublett, C. (2022). The access-equity dualism of online learning in community colleges. *New Directions for Community Colleges, 2022*(198), 25–36. https://doi.org/10.1002/cc.20508

Thompson, N., Allan, J., Carverhill, P. A., Cox, G. R., Davies, B., Doka, K., et al. (2016). The case for a sociology of dying, death, and bereavement. *Death Studies, 40*(3), 172–181. https://doi.org/10.1080/07481187.2015.1109377

Tindall, I., Fu, K. W., Tremayne, K., & Curtis, G. J. (2021). Can negative emotions increase students' plagiarism and cheating? *International Journal for Educational Integrity, 17*(25), 1–16. https://doi.org/10.1007/s40979-021-00093-7

Tuck, E., & Yang, K. W. (2012). Decolonization is not a metaphor. *Indigeneity, Education & Society, 1*(1), 1–40. https://jps.library.utoronto.ca/index.php/des

Turner, R. J., Wheaton, B., & Lloyd, D. A. (1995). The epidemiology of social stress. *American Sociological Review, 60*(1), 104–125. https://doi.org/10.2307/2096348

Verhoef, A. H., Fourie, M., Janse van Rensburg, Z., Louw, H., & Erasmus, M. (2022). The enhancement of academic integrity through a community of practice at the North-West University, South Africa. *International Journal for Educational Integrity, 18*(1), 18. https://doi.org/10.1007/s40979-022-00115-y

Vogt, L., & Eaton, S. E. (2022). Make it someone's job: Documenting the growth of the academic integrity profession through a collection of position postings. *Canadian Perspectives on Academic Integrity, 5*(1), 21–27. https://doi.org/10.11575/cpai.v5i1.74195

Williams, L. (2007). What if (Song). *On West*. UMG Recordings Inc.

Wolbring, G., & Escobedo, M. (2023). Academic coverage of social stressors as experienced by disabled people: A scoping review. *Societies, 13*(9), 211. https://doi.org/10.3390/soc13090211

Corrine D. Ferguson, M.A., is a sociology instructor at Bow Valley College in the School of Arts and Sciences and a member of the Contract Cheating Working Group for the Alberta Council on Academic Integrity. She has been teaching at the college since 2013. Current research focuses on the stress process and academic integrity, including contract cheating. She is particularly interested in involving students as partners in research.

Margaret A. Toye, Ph.D., is an Associate Dean at Bow Valley College in the School of Arts and Sciences and the School of Extended Education. She was a founding steering committee member of the Alberta Council on Academic Integrity. She has engaged in applied research projects on contract cheating, predictors of Practical Nurse graduates' success and failure in the licensing exam, language supports for nursing students, and the relationship between literacy and youth involvement with the law.

Chapter 8
Polytechnic Pinch Points for Academic Integrity: Reflections from Student Conduct Officers

Laura Lucien-Bay, Linden Couteret, and Susan Radke

Abstract Polytechnic institutions take pride in offering hands-on, applied technical training with close ties to industry. Faculty members at polytechnics are often recruited directly from professional workplaces, bringing specific expertise in their area of specialty but often lacking traditional academic or pedagogical backgrounds. A teaching background may be an added bonus for some, but the reality is that many faculty members will not have taught in a classroom before. As such, academic integrity and issues of academic misconduct may not be top of mind for faculty in a polytechnic space. Through the collected reflections of two Western Canadian polytechnic student conduct teams, this chapter identifies five pinch points faced by faculty and staff in polytechnics when addressing academic integrity. The pinch points are (1) the wide spectrum of the student experience, (2) the diverse range of assessments, (3) the impact of connection to industry, (4) shortages of time and disconnected spaces, and (5) the role of regulatory bodies. The chapter offers practical recommendations for each pinch point to help polytechnic faculty and staff uphold academic integrity and effectively address these challenges.

Keywords Academic integrity · Polytechnic · Faculty support · College instructor · Work-integrated learning · Vocational training · Integrity principles · Student conduct · Canada

Polytechnics Canada (n.d.) describes Canadian polytechnic institutions as learning environments where students will prepare to meet the needs of the workforce through applied learning experiences with industry professionals. As a result,

L. Lucien-Bay (✉) · L. Couteret
University of Alberta, Edmonton, AB, Canada
e-mail: Lucienba@ualberta.ca; couteret@ualberta.ca

S. Radke
Southern Alberta Institute of Technology (SAIT), Calgary, AB, Canada
e-mail: susan.radke@sait.ca

L. Vogt, B. M. Stoesz (eds.), *Academic Integrity in Vocational and Polytechnic Education*, Ethics and Integrity in Educational Contexts 8,
https://doi.org/10.1007/978-3-031-90091-4_8

polytechnic faculty hold professional expertise in their area of teaching and often transition into teaching directly from industry. While some faculty may have an academic background or experience in teaching, many will not have training as instructors and have never taught formal classes with course outlines, learning outcomes, structured learning activities, and assessments. Polytechnic assessments are often applied demonstrations or other hands-on applications of the course learning objectives. Recognizing potential academic integrity breaches is a faculty responsibility that brings unique challenges in this context.

The authors of this chapter have worked as student conduct officers in two polytechnic institutions in Western Canada. Although there are variations in the job descriptions of student conduct officers, these positions generally support faculty and academic managers with concerns about academic misconduct and provide students with opportunities to proactively seek information, support, and guidance to avoid issues of academic misconduct. Both institutions serve large urban centers and surrounding rural areas, but there is no affiliation between our institutions. Through provincial and national professional development and community-building activities, we have come to know each other and exchange ideas that support our work. Our discussions have uncovered a shared observation that for many faculty in polytechnic institutions, academic integrity and academic misconduct take a back seat to the subject matter expertise with which the faculty member is more familiar. This may result in faculty being unsure how to address concerns about academic integrity with the student. We have learned that the nature of hands-on learning may change what academic integrity, and therefore academic misconduct, looks like in polytechnic education. While academic integrity at a polytechnic includes proper source acknowledgment and may require referencing and citations, it also encompasses how students represent their abilities in science labs, testing facilities, trades workshops, small interactive classrooms, and external work-integrated learning sites.

Structure of Student Conduct Teams

It is important to note that the student conduct teams are structured and operate differently at our two polytechnic institutions. Within our broader academic integrity community, we have seen that when polytechnics have student conduct teams, they reflect their particular institutions' context and culture, resulting in a wide range of responsibilities and processes. Some institutions have multiple teams, each addressing different student behaviour types. For example, the office dedicated to managing complaints of gender-based and sexual violence may be separate from the office dedicated to managing academic misconduct. Conversely, other institutions have central offices responsible for all the wide-ranging areas of student behaviour, both academic and non-academic. Furthermore, other members of our broader academic integrity community work in polytechnics without any form of student conduct office at all.

Instructors at both our institutions are the first response to academic misconduct allegations; however, each institution has a unique way of handling academic misconduct issues. At one institution, instructors must meet with the students involved and seek their explanation as to what occurred. The instructor balances the student's account with available evidence to ascertain whether academic integrity principles have been violated. If the instructor determines the principles have been violated, they can apply a reasonable sanction following the institutional policy. In more escalated cases, such as where a student has had previous academic misconduct concerns, school leadership will address the case directly and determine higher-level sanctions in line with institutional policy. The centralized academic integrity office supports instructors with administrative tasks related to policy and procedures, offers guidance through consultations and standardized templates, and maintains institutional records. The instructor holds primary decision-making power in this system, and the centralized office provides administrative support.

At the other institution, academic schools and programs are similarly responsible for the day-to-day management of academic misconduct but with a slightly different flow. Instructors flag potential misconduct in their courses and gather the evidence, and then the academic chair and/or other academic managers hold hearings, make decisions, and impose sanctions. The student conduct office supports the academic schools in these processes by developing and maintaining standard hearing notices and decision letter templates, providing advice on challenging situations, entering student misconduct incidents into the student database, and tracking and analyzing statistics for the institution. This institution follows a more quasi-judicial process in addressing student academic misconduct and uses the centralized office as a source of expertise and advice.

Both academic conduct teams work closely with other internal institutional partners, such as the Office of the Registrar, the Library, the Ombudsperson, the Teaching and Learning Centre, and academic leadership. Working together to reflect on our experiences, we discovered that although our offices operate differently, the academic misconduct experiences and pinch points, or areas of difficulty, in our two Canadian polytechnics are similar. Through our collected reflections, this chapter explores five pinch points in polytechnic education and offers recommendations from our practice to support faculty. We recognize that these issues may present similarly in other higher education sectors and foresee that our advice will apply to a wider audience.

Pinch Point 1: The Wide Spectrum of the Student Experience

Polytechnics in Canada generally offer a wide spectrum of specializations, programs, and credentials, including high school upgrading, English language learning, accredited trades and journeyperson training, as well as certificates, diplomas, and degrees in a broad range of technical and applied fields. The plethora of educational choices makes for a vibrant student body, and this is an aspect of polytechnic education to be celebrated. Entrance requirements include the completion of relevant high school courses with a minimum grade; however, students are often admitted by the order of their application rather than a competitive selection process, which democratizes the admission process and increases opportunities for prospective students (Dennison & Gallagher, 1986). However, this also presents a unique challenge for educators, as students come from a wide range of educational backgrounds. For example, some students will have a firm understanding of the importance of integrating and referencing source materials, while others may have little or no previous experience with these requirements and be expected to learn them during their time in post-secondary education (Peters & Cadieux, 2019).

To ensure that all students have the necessary understanding of academic integrity, instructors often need to create resources and tools for student education. A broad range of academic integrity resources can take instructors time to curate, promote and, in some cases, create. Some institutions assign academic integrity development to specific roles, such as academic integrity specialists to promote academic integrity values, dedicated librarians to teach citation, and academic coaches to help students organize their study strategies. Others may have supplemental materials or tutorials available for instructors to use in order to promote good academic conduct. Regardless of the institution's particular approach, an instructor should know what tools are available and build them into their practice—a task that can take some time, depending on the nature and extent of the institution's resources. Clear expectations for academic integrity are important so students understand how to make good decisions and what constitutes academic misconduct (Christensen Hughes & McCabe, 2006a).

Our Response

We encourage instructors and institutions to raise awareness about academic integrity with specific and clear discussions about what is expected in courses and related professional industries. This includes sharing information on relevant resources, creating clear assessment expectations, using regular reminders of requirements, and modelling what is expected. The following are examples of good practices for instructors:

- Include statements on academic integrity as well as links to on-campus academic support in the standard syllabus. Review this part of the syllabus in class, as students may not read it on their own.
- Remind students of the resources available to help them develop academic skills and their responsibility to access those resources regularly throughout the term to reinforce acting with integrity when it is most relevant.
- State academic integrity expectations and consequences of violating the principles of academic integrity using a positive and supportive tone. We advise instructors to avoid creating a fear-based response and focus instead on

demonstrating positive characteristics. The following list illustrates actions that instructors can take to provide a supportive tone toward academic integrity:

  - Use boilerplate statements on assessments to ensure a consistent format and placement of messaging. Examples of formats include charts of what is allowed or not allowed (e.g. George Washington University, n.d.), scales of how specific tools can be used, such as the artificial intelligence assessment scale proposed by Perkins et al. (2024), and checklists (e.g. University of Waterloo, n.d.) which has clearly defined boundaries for the use of AI in the specific assessment, such as no AI allowed at all, AI-assisted idea generation or AI-assisted editing is permitted, for example.
  - Recognize that students may not read these boilerplate statements so instructors should review the expectations in class and provide an opportunity for students to seek clarification. Ashworth et al. (1997) discuss the value of positively framing discussions about academic integrity and providing concrete examples of what is desired and what is unacceptable.

- Model academic integrity expectations through proper citation of resources used in lectures and class materials, using the same system expected of students. Modelling demonstrates that instructors and staff value academic integrity and are held to the same academic integrity expectations, and this provides students with examples to learn from.
- Share experiences of what integrity looks like in industry and relate these values to the International Center for Academic Integrity ICAI (2021) fundamental values of academic integrity: honesty, trust, fairness, respect, responsibility, and courage.
- Integrate institutional resources such as academic integrity courses, modules, and websites, as well as external resources such as Open Educational Resources on academic integrity.
- Collaborate with experts on campus who can provide guidance, such as research librarians and academic integrity specialists.

Pinch Point 2: The Diverse Range of Assessments

Education that applies hands-on, experiential learning methods will require hands-on, experiential assessments. A significant component of a polytechnic credential is that students are assessed on their ability to demonstrate skills; therefore, a final assessment might require students to operate equipment, create a finished product, devise a solution to an industry problem, or exhibit particular competencies at a work site. In these experiential assessments, academic integrity breaches can present differently than in exams and term papers. For example, our offices have supported cases where students submitted falsified signatures from onsite supervisors after work-integrated learning experiences; students misappropriated designs created by industry partners and even secretly brought in pre-made foods for final culinary arts assessments. The wide range of assessment methods means the instructor must be aware of what academic integrity looks like (or tastes like!) in each assessment and be prepared to verify when a breach of academic integrity has occurred.

Our Response

By its very nature, this pinch point is hard to address through a single response. Our offices have found that when instructors set clear expectations in advance, as described in pinch point one, students are better able to avoid academic misconduct. If academic misconduct does occur, we recommend that instructors refer to the set expectations to discuss the issue and the next steps with the student.

When communicating academic integrity expectations about an assessment, consider the nature of the assessment and ways that students must demonstrate academic integrity. Do not assume that students know what the expectations are in each situation. It is important to illustrate how to act with academic integrity for each assessment so that students have a clear understanding of what is required. The following examples offer some key areas to consider:

- If students must work collaboratively, describe what that collaboration should look like in this particular situation. This will avoid confusion about what constitutes collusion in different courses and on specific assignments, which can contribute to academic misconduct (Barrett & Cox, 2005). For example, is each student expected to do part of the work, and the group will synthesize those parts? Or is the group expected to work together at each point? Is collaboration allowed only between members of the group, or is collaboration also allowed between different groups?
  - Figure 8.1 provides an example of a tool that can be used to talk about collaboration expectations.
- If students are being assessed on their ability to create something, be clear about what tools and resources they can use. If the work is to be completed in a timed classroom setting, what can they bring with them, and how much work (if any) can be done in advance?
  - For example, can students bring a plan or template with them? If so, does it need to be a template they have created, or can they use a pre-existing plan or template created in class? Are they expected to bring their own materials or to use only what is provided? Can they review examples or other resources during the activity, or do they have to work from memory? Can students interact during the assessment? If so, what type of interaction is allowed, and are there limits to when and how frequently?

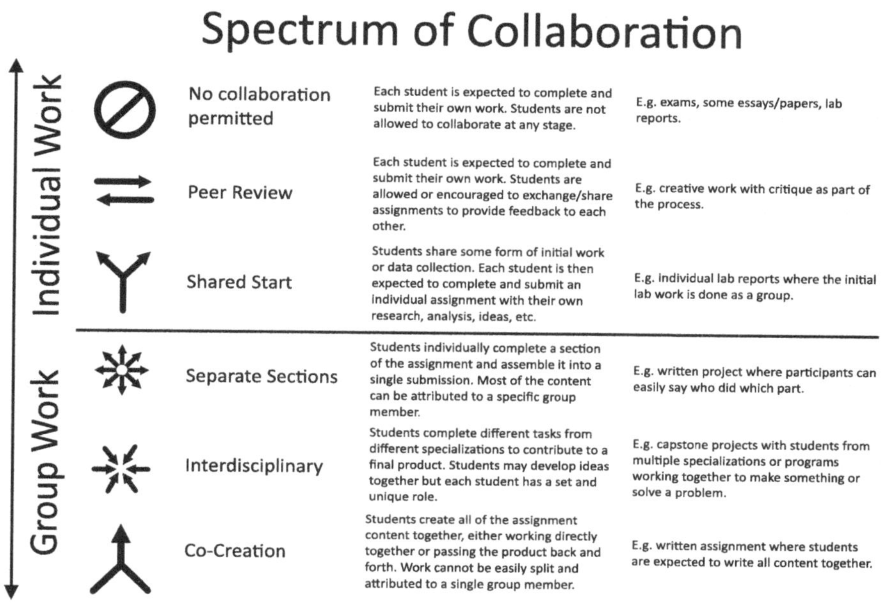

Fig. 8.1 Spectrum of Collaboration. *Note:* Diagram provides a short summary and examples of different levels and types of collaboration on assessments. Created by Linden Couteret

- If students are assessed at the work-integrated learning site, consider how feedback will be gathered. What role does the student have in that process? Do they gather the feedback and submit it to the instructor, or does the site supervisor submit it directly? Does the supervisor have to sign off on the feedback? What happens if there are inconsistencies?
  - For example, if there are any discrepancies or concerns with information presented by the student, the instructor should follow up with the supervisor.
- If students are being assessed on multi-step projects, consider asking students to submit drafts or a record of their process in stages. This helps the instructor identify issues early on and lends evidence that the student has engaged in a process to create a final product.

Pinch Point 3: The Impact of Connection to Industry

Research states that faculty at Canadian post-secondary institutions, such as universities, may ignore academic misconduct due to lack of evidence, low administrative support, limited time, and the perceived low risk of the offence (Christensen Hughes & McCabe, 2006b). However, in the polytechnic setting, we have observed additional factors that result in ignoring academic integrity breach issues related to industry connections. For one, the impact of an allegation of academic misconduct can bring the instructor a nuanced burden in their relationship with industry partners. For example, many instructors take pride in providing references to students going into their industry, whether for a work-integrated learning experience or employment upon graduation. This is particularly true in smaller industries where the instructor is well-known and widely regarded. An instructor may feel pressured not to pursue or even to ignore an academic misconduct concern because they do not want to tarnish their relationship with a student who they believe will, despite the academic misconduct and with a bit of coaching, make an excellent colleague in the workplace.

Another instance where polytechnic instructors may be hesitant to proceed with academic misconduct allegations is when it conflicts with how their industry views and addresses similar issues. In some cases, the academic requirements for integrity, such as the citation of source material, are greater than those in industry. An instructor may be reluctant to pursue the issue for fear that the polytechnic will take a heavy-handed approach to an issue with little relevance in the workplace. As a result of the issues above, an instructor acting with good intentions may delay or even mute responses to academic misconduct.

In a reverse scenario, an instructor may issue a more severe consequence for a student's action than the institution requires. This might be because the instructor knows that this action would warrant dismissal in the workplace but that the institution views the issue as part of learning and, therefore, requires less stringent consequences. When instructors seek to determine their own consequences, which are different from the institutional policy, there is a problem of inconsistency, and students end up uncertain about the expectations.

Our Response

We advise instructors to remind themselves that in this role, they are educators first, supporting students on a learning journey rather than managing employees in the workforce. Good teaching should co-exist with other professional practices; however, teaching and learning principles and institutional policy must take precedence. Focusing the conversation on policy rather than personal expectations can also help an instructor manage the delicate balance of being a representative of industry and an advocate for academic integrity. For example, an instructor can use starter statements such as 'the rules of our institution require that I follow up with you' when meeting with students to help re-align the student's conduct to the expectations of the educational environment.

Instructors may also benefit from learning about restorative practices for addressing academic misconduct. This approach centres on repairing relationships and addressing the harm caused by academic integrity breaches. Restorative practices can be delivered in tandem with institutional academic integrity policy, even when restorative approaches are not explicitly part of the policy or procedure. Instructors who focus on repairing relationships offer a meaningful way to support academic integrity and develop students as ethical professionals. Further details on restorative approaches for polytechnic instructors are outlined by Fukushima and Couteret (2025).

Pinch Point 4: Shortages of Time and Disconnected Spaces

Polytechnics offer a range of program lengths, including very short programs. For example, a trades apprentice may only be in classes for six to eight weeks of intensive instruction and lab work per year. Short programs with tight timelines and fast-paced schedules increase the demands on instructors and may discourage them from dealing with academic misconduct. Alberta Colleges and Institutes Faculties Association's (ACIFA, 2023) position paper on academic integrity underscores how heavy workloads among college faculty limit time for learning and implementing strategies to uphold institutional standards of academic integrity.

Furthermore, limited time and resources to address academic misconduct can be exacerbated by campus size and location. Due to the size of the geographical area served or the unique space needs of specialized programs, polytechnic institutions may have satellite campuses, and staff in these locations may be less connected to the resources available at the main campus (Helengwa, 2014). For example, academic integrity professionals or librarians may be located at the main campus and unable to devote much time to the smaller satellite campuses. As a result, there can be limited knowledge of the academic integrity support available for faculty and students at satellite campuses, and accessing these services can be more challenging.

Our Response

A structured approach to academic integrity policy and procedure can support instructors when navigating issues of academic misconduct in tight timelines or across disconnected campuses. A structured approach includes proactive and ongoing activities that keep academic integrity as a thread when instructors and staff prepare for a new cohort, intake, or semester. These activities can include:

- Review the institutional policy and procedure regularly to ensure a clearly defined scope for academic misconduct and to detail the steps needed to address misconduct allegations.
- Provide a resource library with templates and drafts for instructors to communicate academic misconduct concerns and complete required administrative tasks to decrease the workload and siloed responses.
- Build a network of instructors with expert knowledge at satellite campuses who can provide peer support on addressing academic misconduct.

- Advocate for the institution to address service gaps that impact faculty's ability to follow up with student academic integrity breach concerns, such as lack of in-person library and academic support services on satellite campuses,
- Introduce faculty to support services ahead of each new intake to ensure faculty know who can help educate students on academic integrity and address situations of academic misconduct. This should include the conduct office, the library, and others involved in promoting academic integrity.
- Explore creative ways of providing timely and accessible support for students and instructors, such as virtual consultations and pop-up sessions at satellite locations.

Pinch Point 5: The Role of Regulatory Bodies

Polytechnic educators are driven not just by institutional responsibilities but also by the responsibilities outlined by industry regulators. These regulators can include provincial apprenticeship boards, health service governing bodies, oil and gas industry oversight bodies, engineering accreditors, federal regulators such as Transport Canada, and other external certification providers, to name but a few. The regulator's role is to ensure that the program material being taught aligns with current industry standards and legislative requirements. Regulators will review the course material and learning equipment regularly. In some programs, regulators will also administer their own certifications, licensing, and/or assessments to verify that new graduates coming into industry are up-to-date and meet industry standards.

Regulators can have their own requirements of academic conduct and how to handle breaches of academic integrity, which may differ from the institution's requirements and processes. For example, the regulating body may use a punitive approach, such as requiring a grade reduction or removing a credential when responding to academic misconduct to meet its standards. However, an academic institution may use an educational approach, such as requiring the student to complete a remedial academic integrity course or a reflective activity, rather than using a punitive measure, such as lowering a grade or rescinding a credential. This can cause confusion for instructors as they navigate their responsibilities to industry and the institution.

Additionally, regulatory and accreditation requirements can significantly influence a program's curriculum, and there is a risk of increasing academic misconduct among students. When the learning outcomes, teaching style, and assessment choices of a particular course are geared toward meeting industry regulations and government legislation, this can create high-stakes situations for students. For example, an instructor may rely on multiple choice exams to prepare students for an upcoming regulator exam rather than using alternative, low-pressure forms of assessment. A student's failure to successfully complete these built-in certifications can result in the student failing the course or even the entire program. This type of high-stakes assessment puts increased pressure on students and creates a heightened risk of academic misconduct, as students may feel pressure to breach the institution's academic integrity policy to complete these courses successfully.

Our Response

If there is a conflict between regulation standards and institution policy, an instructor can:

- Refer to the institutional academic integrity policy and procedure to determine if the role of professional codes of conduct is described. In some cases, the institution will determine that breaches of professional codes of conduct will be pursued under the professional policy rather than the institutional policy.
- Consult with their academic leader as this problem may have been identified before, and a suitable middle-ground may have been identified. For example, if an instructor has opted for a group project to assess the achievement of a learning outcome but the regulator's assessment is a multiple-choice exam, the instructor may opt for multiple-choice quizzes as a formative assessment to give students the opportunity to practice meeting the regulator's assessment expectations, without compromising the deeper assessment of learning as required for their course.
- Discuss the regulatory standards and expectations of the regulatory body with students and highlight where there is a connection between academic integrity in coursework and the expectations of the regulatory body and where the two differ.
- Advocate for institutional leadership to connect with regulators and address discrepancies in assessment methods and policy application.

Conclusion

As student conduct officers, we have worked with polytechnic instructors and students to handle a wide variety of difficult situations involving academic misconduct. In this chapter, we have shared five pinch points for academic integrity in polytechnics and offered responses to address these issues based on our practice. As we conclude our reflections, we have identified the following four key benefits for instructors that resonate from our responses to the pinch points:

1. Instructors benefit from receiving support in the transition from industry to education.
2. Instructors benefit from knowing the institutional academic integrity policy.
3. Instructors benefit from communicating clearly with students.
4. Instructors benefit from seeking support to respond to academic integrity issues.

All four require proactive efforts from instructors, support staff, and academic leaders, and in our experience, this creates a more supportive environment where both instructors and students will be more likely to act with academic integrity. By proactively reviewing policies and procedures, polytechnic staff are better equipped to handle issues that arise with fairness and consistency. Proactive communication ensures that students understand academic expectations and boundaries, reducing the likelihood of academic misconduct. Finally, seeking support from colleagues and institutional resources can reduce stress and strengthen instructors' ability to carry out their responsibilities in upholding the academic integrity policy. These benefits extend to students who will be better prepared to navigate the complexities of academic integrity and demonstrate honesty and responsibility in their academic work, resulting in increased confidence to handle ethical challenges in their future careers.

References

Alberta Colleges and Institutes Faculties Association (ACIFA). (2023). *ACIFA's position on academic integrity* (Position paper). https://www.acifa.ca/_files/ugd/16fcf0_c98c6fb9f9b-c4e4880a9f875d754eb4e.pdf

Ashworth, P., Bannister, P., & Thorne, P. (1997). Guilty in whose eyes? University students' perceptions of cheating and plagiarism in academic work and assessment. *Studies in Higher Education, 22*(2). https://doi.org/10.1080/03075079712331381034

Barrett, R., & Cox, A. L. (2005). 'At least they're learning something': The hazy line between collaboration and collusion. *Assessment & Evaluation in Higher Education, 30*(2), 107–122. https://doi.org/10.1080/0260293042000264226

Christensen Hughes, J. M., & McCabe, D. L. (2006a). Academic misconduct within higher education in Canada. *Canadian Journal of Higher Education, 36*(2), 1–21. https://doi.org/10.47678/cjhe.v36i2.183537

Christensen Hughes, J. M., & McCabe, D. L. (2006b). Understanding academic misconduct. *Canadian Journal of Higher Education, 36*(1), 49–63. https://doi.org/10.47678/cjhe.v36i1.183525

Dennison, J. D., & Gallagher, P. (1986). Characteristics of Canada's community colleges. In *Canada's community colleges: A critical analysis*. UBC Press.

Fukushima, A., & Couteret, L. (2025). Building relational culture in polytechnic education through restorative approaches to academic integrity. In L. Vogt & B. M. Stoesz (Eds.), *Academic integrity in vocational and polytechnic education*. Springer.

George Washington University. (n.d.) *Promoting academic integrity*. Retrieved May 28, 2024, from https://studentconduct.gwu.edu/promoting-academic-integrity

Helengwa, D. (2014). The challenges of working and studying at a satellite campus of a university: A case of the Durban University of Technology. *Mediterranean Journal of Social Sciences, 5*(27) https://www.richtmann.org/journal/index.php/mjss/article/view/5127

International Center for Academic Integrity [ICAI]. (2021). *The fundamental values of academic integrity* (3rd ed.). ICAI. https://academicintegrity.org/images/pdfs/20019_ICAI-Fundamental-Values_R12.pdf

Perkins, M., Furze, L., Roe, J., & MacVaugh, J. (2024). The artificial intelligence assessment scale (AIAS): A framework for ethical integration of generative AI in educational assessment. *Journal of University Teaching and Learning Practice, 21*(6) http://open-publishing.org/journals/index.php/jutlp/article/download/810/769

Peters, M., & Cadieux, A. (2019). Are Canadian professors teaching the skills and knowledge students need to prevent plagiarism? *International Journal for Educational Integrity, 15*(10). https://doi.org/10.1007/s40979-019-0047-z

Polytechnics Canada. (n.d.). *Polytechnics at work.* https://polytechnicscanada.ca/wp-content/uploads/2024/05/Polytechnics-at-work_2024.pdf

University of Waterloo. (n.d.) *Academic integrity checklist.* https://uwaterloo.ca/academic-integrity/sites/default/files/uploads/documents/academic_integrity_checklist_0.pdf

Laura Lucien-Bay (she/her) is a leader who has coordinated student conduct offices at two large post-secondary institutes in Alberta, Canada. She prides herself on trauma-informed decisions regarding student conduct, as well as developing policy that balances institutional directives, legislation and humanity. She has been an administrator in post-secondary education for over twenty years, leading in a variety of roles, from disability and accessibility services to teaching services and student conduct. She has a Master's degree in Post- Colonial Studies from the University of London, as well as certificates in Crisis Management and Adult Education.

Linden Couteret (they/them) is a student conduct administrator with a background in teaching, student learning success, and disability services. They have a Master's degree in education focused on developing student metacognition, a certificate in restorative justice, and a Bachelor's degree in secondary education.

Susan Radke (she/her) has held a variety of roles in a polytechnic environment for thirty years, including instructing, managing academic programs as an academic chair, managing policies and procedures for the institution, and leading the student conduct office. She has a B.A. from the University of Alberta, an LL.B. from Dalhousie University, and an LL.M from the University of Alberta, as well as a Master's Certificate in Project Management from York University/University of Lethbridge.

Chapter 9
Advocating for Fair Compensation and Employment Equity: Reducing the Burden of Academic Integrity and Misconduct Work in Colleges and Polytechnics

Brooklin Schneider

Abstract Instructors in colleges and polytechnics, especially those with employment history in the trades and/or industry with limited training in teaching and other academic practices, face challenges in identifying and reporting academic misconduct. The formal processes for documenting and addressing such incidents can be cumbersome and time-consuming, exacerbating the already substantial workload associated with teaching and service duties. In addition, dealing with cases of academic misconduct is often stressful, resulting in burnout. The workload and emotional burden may be particularly heavy for sessional instructors who (1) are paid a limited number of hours within a set time period that may not adequately account for the extra work hours necessary to deal appropriately with cases of academic misconduct and/or (2) fear the repercussions for reporting from students and administrators. Permanent faculty, by comparison, do not typically receive extra compensation, but hold ongoing positions that include paid holidays. Stable employment with increased benefits is helpful to offset some of the emotional and unremunerated labour that accompanies academic misconduct detection and reporting.

This chapter explores the idea that the promotion of academic integrity and the prevention, detection, and reporting of academic misconduct are labour issues that impact the employment equity of instructors teaching in colleges and polytechnics. This chapter also calls for systemic institutional changes to support instructors in three primary areas—time and compensation, administrative support, and professional development. To support instructors more fully, academic integrity and academic misconduct work must be included within formal workload calculations and collective agreements to ensure fair compensation. In addition, opportunities and funding must be granted to instructors so they can engage in professional develop-

B. Schneider (✉)
NorQuest College, Edmonton, AB, Canada
e-mail: Brooklin.Schneider@NorQuest.ca

L. Vogt, B. M. Stoesz (eds.), *Academic Integrity in Vocational and Polytechnic Education*, Ethics and Integrity in Educational Contexts 8,
https://doi.org/10.1007/978-3-031-90091-4_9

ment activities to become better equipped to promote academic integrity and deal with academic misconduct. Collective advocacy may be required to spark systemic changes within colleges and polytechnics and to collective agreements to mitigate the burdens associated with academic misconduct. By fostering more supportive and equitable environments, institutions can better sustain cultures of integrity and enhance the well-being of their instructors.

Keywords Academic integrity · Academic misconduct reporting · College faculty workload · Faculty labour · College instructors · Employment equity · Advocacy · Community support

College and polytechnic faculty are deeply committed to upholding the values of academic integrity. One aspect of this commitment, however, is the corresponding costs of unremunerated work hours, emotional labour, and burnout associated with dealing with academic misconduct. Many college and polytechnic instructors have extensive expertise and experience as a result of working in their discipline, trade, and/or industry. They may, however, have little formal training in the professional practice of teaching, including the promotion of academic integrity or skills for identifying, documenting, and reporting plagiarism or other forms of cheating. This reduced knowledge and skills can result in a lack of confidence and feeling overwhelmed and overworked, which can be particularly problematic for sessional instructors who may also fear repercussions from the administration because of student complaints and negative course evaluations (Crossman, 2019, 2021). Academic misconduct is, therefore, an important workplace issue.

This chapter examines the workload conditions associated with promoting academic integrity and preventing, detecting, and reporting academic misconduct, which exacerbate burnout. The chapter concludes with a discussion of the third space advocacy work and communities of practice (CoPs) in provincial and national contexts. These shared, supportive environments can effect change to reduce the workload and manage emotional burdens and burnout associated with academic integrity and misconduct work for instructors across the employment spectrum at colleges and polytechnics.

The Academic Misconduct Burden

The commitment to upholding and enforcing academic integrity is more than just a professional responsibility – it is an ethical imperative to safeguard an institution's reputation, credentials, and society (Dawson, 2021). Openo (2024) drew attention to two highly publicized cases that called the integrity of engineering education and engineers' work into question. On March 11, 2018, CBC News exposed cheating by engineering students at the University of Regina. Just days later, a pedestrian bridge collapsed at Florida International University, resulting in multiple deaths and injuries. An investigation revealed structural issues originating from the construction of the bridge. Openo drew connections between these two incidents, stressing the very real safety consequences of academic dishonesty, stating that "...the current volume of academic integrity violations are the screaming cracks in the concrete for post-secondary institutions. They are screaming that something is seriously wrong, and solutions must be found because it is quite possible that lives are at stake" (p. 217). Students graduating with credentials via unethical and dishonest academic practices then join the workforce as incompetent but hidden in plain sight.

Discussions regarding the risks of academic misconduct to public safety often center on university programs with clear and direct links to health and human safety: It is an oft-repeated idiom that nobody wants a surgeon who cheated on their exams performing open-heart surgery. And yet, the stakes can be just as high for students training in college, vocational, and polytechnic programs to work in such fields as avionics, health care, electrical trades, and engineering technology. These are the workers "who keep our systems running safely and smoothly, making academic integrity a safety issue" (Vogt, personal communication, 2023, March 29). Upholding academic integrity and dealing with academic misconduct are necessary to safeguard the public. Hamilton and Wolsky (2022), however, identified four barriers to reporting allegations of academic misconduct at Canadian colleges and polytechnics:

1. Increased time and workload involved in reporting incidents,
2. perceived lack of support from senior administration,
3. lack of awareness of institutional academic integrity policies, and
4. perceived threats if faculty report incidents.

Academic integrity and academic misconduct labour are often invisible, taken for granted and considered to be outside of the traditional workload of instructors. Where is this work located? Who does this work? What are its associated job duties?

Those who do work related to academic integrity and academic misconduct operate in a "liminal Third Space" (Ahuna et al., 2023, p. 76), because it is not described in job descriptions, does not fit neatly into organizational structures, and is difficult to categorize despite being the foundation of all aspects of higher education. Moreover, "those ... who do the work ... [often struggle] for recognition and [feel] overlooked, misunderstood, or undervalued" (Ahuna et al., 2023, p. 76). This third space lens echoes calls from academic integrity scholar-practitioners to render this labour visible by appropriately balancing instructor workloads to allow for time- and resource-intensive demands of academic integrity work, and to legitimize and formalize it into official positions on the organization chart (i.e., see #makeitsomeonesjob, Murdoch & Ellis, 2021, in Vogt & Eaton, 2022). In addition, academic integrity work must be "compensated in a manner that demonstrates the work is valued by the institution" and accompanied by "increased professional learning and training opportunities" (Vogt & Eaton, 2022, p. 21). Without these formal positions and appropriate compensation, higher education professionals and the work that they do will continue to be devalued and de-prioritized (Vogt & Eaton, 2022).

The invisibility of the work and barriers to upholding academic integrity values not only disincentivize reporting but also contribute to the emotional and physiological toll of upholding a culture of academic integrity. In this chapter, I argue that bargaining tables, where instructors and employers negotiate the terms of collective agreements, are part of the *Third Space of Academic Integrity*. Provincial bodies and CoPs that lobby on behalf of and support instructors, respectively, and render academic integrity labour visible also operate in this third space.

Time and Workload

It is widely acknowledged that postsecondary instructors (including sessional instructors) work more hours than are captured in their standard workload assignments. A standard workload for a full-time instructor at a college or polytechnic in Alberta (and other provinces in Canada) ranges from 80 to 100% teaching and 0 to 20% service (e.g., committee work, curriculum development or revision, research, union business, and other duties as assigned). Any work that is not considered part of the contracted work of teaching or grading (which can include identifying, investigating, documenting, and reporting academic misconduct) may be captured amorphously in the 20% service language of instructor job descriptions. More often than not, however, the work associated with academic misconduct is invisible and is not accounted for in the language of faculty union collective agreements. In addition, dealing with academic misconduct often requires more time than is remunerated.

At some colleges and polytechnics, instructors may take on administrative roles and may act as the decision-makers in academic misconduct cases and appeals — this additional commitment may lead to working overtime hours that are also unremunerated. Not only do they shoulder the responsibility of investigating and reporting cases from students in their programs or department areas, but they must adjudicate the cases from their own classes. The caseload can be heavy at peak times (such as the end of a semester), resulting in backlogs of several weeks to months to process academic misconduct cases. The backlogs are not good for students nor for instructors and administrators who find themselves physically and emotionally burned out (Openo & Robinson, 2021). Unfortunately, this overload has been normalized.

Advocating for Change in Working Conditions

In Alberta, post-secondary institutions operate under two provincial legislative acts: the Post-Secondary Learning Act (PSLA, 2003, c P-19.5) and the Labour Relations Code (2000, c L-1). The PSLA governs the composition and operations of post-secondary institutions in Alberta, defining the governance structures, roles and responsibilities for boards of governors, academic councils or senates, student affairs, and other official bodies. The Labour Relations Code stipulates employment conditions and sets out the requirements for collective bargaining, job action (strikes and lockouts), and other union-related matters. Post-secondary institutions and instructors employed at them must negotiate the two jurisdictional contexts: education and labour relations.

The Alberta Colleges and Institutes Faculties Association (ACIFA) supports negotiations in these contexts and helps to navigate the complex challenges brought about by working in post-secondary institutions with ever-increasing workload demands and complex teaching and learning practices. ACIFA is a federated association representing the faculty associations (unions) consisting of instructors (both teaching and non-teaching academic staff) from 10 colleges, two polytechnic institutes, and one university, ranging in size. Credentials are varied, with a focus on short certificate programs of 4 weeks to 8 months, to 2-year diploma programs, 3-year applied degrees, and 4-year degrees. Academic and vocational programming range from apprenticeship training to academic upgrading (for secondary school equivalencies), to accredited professional (e.g., nursing, health care aid, social work) and university programming, with a focus on vocational training. In addition to labour relations support, ACIFA acts on behalf of its members in the following ways:

- Serves as a provincial voice for faculty members,
- Establishes and manages a province-wide legal defence fund,
- Advocates for quality post-secondary education,
- Facilitates information exchange between Faculty Association Presidents,
- Draws from the experiences of associations in other provinces, and
- Represents members at the Canadian Association of University Teachers (CAUT).

Workload is a perennial bargaining priority for ACIFA and its member associations. Collective agreements (or contracts) stipulate the terms and conditions of employment, including working conditions, wages and benefits, for instructors employed at their respective institutions. Collective agreements also define instructor job duties, including teaching preparation, curriculum work (course and program development), student assessment, student assistance, coaching/mentoring and other forms of leadership or administration, and may include language pertaining to research/scholarship, service work (e.g., committees), and other professional responsibilities (e.g., maintaining professional licenses for accredited/regulated professions). Collective agreements offer high-level frameworks but are not exhaustive; however, implicit in the name—*collective* agreement—instructors and the institutional administration share responsibility for the work of the institution.

Quantifying, standardizing, and codifying all workloads into collective agreements is complex. Instructional hours (or student contact hours) are more easily quantified, ranging from 15 to 22 h of a 40-h work week. To attempt to quantify and codify the remaining hours, different colleges and faculty associations use a variety of calculations based on factors that impact workload. These calculations are also used to ensure equitable workloads across the bargaining unit even when the specific work of individual instructors differs significantly across the institution.

Workloads are often calculated using some variation of *instruction + service*. Therefore, a workload calculation for an instructor teaching in a 2-year social work diploma program based on an 80% teaching (including instruction and preparation) and 20% service workload distribution within a 40-h work week might be:

> 40 h/week = [15 instructional hours + 17 preparation hours] + [8 h service and professional responsibilities]

Recognizing that this calculation oversimplifies and elides the complex workload factors associated with instructor work, some faculty associations use a multiplier that supposedly captures the unquantifiable aspects of instructors' work. These workload distribution calculations are also typically adjusted according to the credential awarded and class size. Instructional hours are commonly higher for instructors teaching in apprenticeship programs or upgrading (secondary) programs and then are incrementally lower up the credential ladder: Diploma/Certificate, Applied Degree, Degree.

Beyond credential level and class size, other factors for determining workload include the type of instruction (lecture, lab, clinical, practicum), mode of delivery (in person, online, blended, hybrid/hyflex), method(s) of assessment, whether the course is new or existing, the instructor's familiarity with the course, and any other factors deemed to impact workload. Promoting academic integrity and identifying and reporting misconduct and the associated workload are not explicitly stated as factors in workload calculations. Students across the credential spectrum engage in academic misconduct. Perhaps this fact is banal so as not to warrant being included as a unique *factor* that impacts workload. This perspective, again, renders academic integrity work invisible. Faculty associations and institutions, therefore, must enshrine academic integrity work in the formal language of collective agreements so that it can be recognized as necessary labour, included in workload calculations so that all instructors (sessional, term, and permanent) are remunerated fulsomely for their work (ACIFA, 2023).

Lack of Administrative Support

Writing from the Australian university perspective, Kane Murdoch (2024, April 2) emphasized the massive, collective, and institutional-level work needed to address academic integrity challenges. He asserted that systemic solutions are needed to address systemic problems:

> [I]f your uni's response to contract cheating is to do nothing and hope academics pick it up, just stop. You need tools and a team to investigate, because you won't get anywhere one by one-, or with-time poor academics doing this as "service". This is a mass issue, across multiple national education sectors. So, get serious. (Para. 9)

All members of educational institutions need to take responsibility for upholding academic integrity. This does not mean that the workload burden on instructors completely disappears; rather the bulk of it shifts to an institutional (or even multi-institutional) level where the resources, supports, and capacity exist to undertake this work. Institutional policies and procedures (and indeed, cultures) that have individual instructors bear this labour obligates instructors to engage in the wrong kind of labour: misdirected, unproductive, and woefully inadequate to address the scale of the problem, but at the high individual cost of instructor welfare and well-being.

Advocating for Increased Administrative Support

Management Recognition is the acknowledgement in a collective agreement that where a specific article or language of the agreement does not address a particular condition of employment (i.e., is silent), then management (i.e., the employer) is the sole authority on that condition (e.g., Government of Alberta & The Alberta Union of Provincial Employees, 2021). As the work of promoting academic integrity and dealing with academic misconduct is not stipulated in collective agreements, the responsibility falls to the employer. This responsibility is often enacted in the form of policies and procedures governing academic integrity work, including guidelines for investigating, reporting, and sanctioning instances of academic misconduct. Note that while academic integrity typically falls under the "Management Recognition" article in collective agreements, the relevant policies and procedures situate this work within the scope of individual instructors' duties to detect, investigate, report, and sanction instances of academic misconduct.

Insufficient Time and Funding for Professional Development

Training instructors and supporting their ongoing professionalization practices are key institutional functions, and this commitment is commonly stated in the collective agreements. Maintaining currency in their discipline and implementing best practices in teaching and learning are clear priorities. Professional development, therefore, is not optional – it is essential to support instructors who are subject matter experts in their disciplines or industries but may not be experts in teaching, learning, curriculum, student assessment, or academic integrity. As frontline workers with students, "Faculty play a crucial role in creating environments that expound and uphold the values of academic integrity" (Hamilton & Wolsky, 2022, p. 467). Finding the time and funding to engage in professional development, however, is challenging when workloads related to dealing with academic misconduct take priority over professional development. Unfortunately, professional development is not accessible for instructors in some employment categories. Sessional instructors typically have the least access, whereas permanent full-time instructors have the most access. Importantly, regardless of employment status or experience level, instructors should be able to engage in professional development opportunities specific to the unique challenges of identifying, investigating, and reporting academic misconduct.

Advocating for Increased Support for Professional Development

Faculty associations must bargain for mandatory academic integrity training as part of a robust onboarding program for new instructors. Relegating this kind of training to professional development means it's not deemed *business critical training* and is instead viewed as a nice-to-have addition to instructor duties. ACIFA (2023) recommends that "Institutions should ensure onboarding practices provide faculty with a full and deep understanding of academic integrity and the resources and supports available to foster a culture of academic integrity" (p. 10). For this recommendation to be feasible, college and polytechnic instructors require the corresponding release time and reduced workloads to complete the onboarding, and academic integrity and misconduct training. Once again, academic integrity work must be considered an important workload factor in workload calculations and embedded into the professional duties, roles, and responsibilities as outlined in collective agreements to ensure sufficient time and remuneration. This advocacy work at bargaining tables would incentivize promoting academic integrity and responding to academic misconduct and attenuate the impact of barriers to fostering cultures of academic integrity.

Annual Conferences

ACIFA hosts an annual spring conference at one of Canada's most picturesque and iconic rocky mountain locales: Banff, Lake Louise, Canmore, or Jasper. For a few days in the mountains, members are invited to present their current research and teaching practices (including academic integrity work) and discuss union issues with colleagues from across the province. The conference is largely funded through members' union dues and offers an important and necessary opportunity for not only professional development but also a retreat for instructors to seek support, be relieved from their regular job duties, and engage in wellness activities (e.g., hiking and enjoying evening entertainment). Outside of the formal conference schedule, it is common for instructors to informally discuss their challenges, including the emotional labour of reporting academic misconduct, and share their complex emotions associated with this work. Walking and talking with colleagues around the beautiful mountain setting can be restorative and allows for cathartic discussions about a particularly challenging academic misconduct case and the workload burden or experiences of burnout. Connecting to this community of peers, in larger group conference or in informal one-on-one settings, affords instructors meaningful and authentic supports in psychologically safe contexts where they can disclose their challenges, have their experiences validated, and share practices and learnings that they can implement upon returning to work at their respective institutions.

Communities of Practice (CoP): Support for Coping with Emotional Labour and Burnout

Like many Canadian provinces, Alberta is geographically large with a distributed population, and colleges and polytechnics are separated by great distances. College and polytechnic instructors may also have unequal access to professional development opportunities, including conferences about teaching, learning, and academic integrity. CoPs, including those that are held virtually, can bridge professional development equity gaps and serve as important support systems for instructors and others involved in the work of academic integrity and academic misconduct. One such thriving no-cost CoP is the weekly virtual *Academic Integrity Hour (currently known as Canada's Academic Integrity Hour)*, founded by Dr. Sarah Elaine Eaton of the University of Calgary in 2020, that originated as a response to the escalating academic misconduct challenges experienced by instructors and administrators in the first half of the COVID-19 pandemic. Highlighting her motivations for creating the CoP, Eaton (2020) described the individualistic nature of academic integrity work at higher education campuses:

> In general, there are very few people on a single campus whose job is dedicated to academic integrity. In most cases, it is one component of a larger professional portfolio. The result is that individuals who work in academic integrity can be and feel isolated and lack opportunities to connect with one another and develop professionally (p. 4).

This CoP, therefore, was established to support higher education professionals, administrators, and faculty, including sessional instructors with little to no professional development funding, at a time of increasing feelings of loneliness and isolation (Eaton, 2020; Eaton et al., 2021).

In 2021, Eaton et al. reflected on the Academic Integrity Hour's first year as a CoP. The authors emphasized feeling validated upon hearing from colleagues across Canada grappling with similar challenges related to academic misconduct detection and reporting, assessment design, student inclusion and equity concerns, and other common issues. Eaton et al. also affirmed the value of the CoP as a psychologically safe space to ask questions, puzzle through issues, and share perspectives with no risks to job security or employment advancement. New learnings and insights can then be shared with their colleagues at their institutions. *Canada's Academic Integrity Hour* has since grown into a thriving community that welcomes postsecondary instructors, researchers, graduate students, administrators, and higher education professionals from across Canada, regardless of educational mandate, location, or size.

The emotional labour and burnout associated with dealing with cases of academic misconduct continue to be important topics of discussion. In March 2024, at

Canada's Academic Integrity Hour, 21 faculty, educational developers, academic integrity researchers, conduct officers, student affairs directors, and graduate students expressed their own emotional labour and burnout and observed these in others. Attendees then discussed strategies to support others through difficult times, including advocating for additional supports for faculty and staff engaged in academic integrity and misconduct work. Helpful supports can include individual supports for the person investigating a suspected case of misconduct, community supports for faculty, staff, and students engaged in the process, and training that specifically provides guidance on navigating the complex issues derived from academic misconduct cases. Attendees also acknowledged the significant role that institutional culture plays in faculty both feeling and being supported to identify, investigate, and report academic misconduct cases.

Discussions with colleagues and the sharing of resources (e.g., ACIFA, 2023) validates that burnout is real. Donna Feledichuk, a senior administrator at the small, rural Portage College in northeast Alberta, emphasized the value of connection and community in the Academic Integrity Hour CoP:

> We often work in silos in post-secondary institutions. Siloed from our peers in other institutions and siloed at times from our colleagues at our own institutions. We frequently work in isolation, feeling our struggles and challenges are all our own. When we have the opportunity to come together as colleagues with similar concerns from different colleges and universities across the nation we see instead our challenges are often similar to others and solutions and lessons learned from our peers can be customized to work at our own institutions (Eaton et al., 2021, pp. 7–8).

Attendees of the CoP provide empathic support and help to create an open venue to share thoughts, feelings, concerns, and resources. The CoP provides a necessary third space in which the challenges associated with the emotional work of academic integrity and academic misconduct within the college or polytechnic context are affirmed.

Academic Integrity Support in Various Regions

From 2020 to 2023, ACIFA's Professional Affairs Committee did their best to respond to and navigate the complex working conditions during the COVID-19 pandemic, which exacerbated existing academic integrity concerns in higher education (see Eaton et al., 2023a). The committee pivoted to a more supportive role, holding meetings to share, brainstorm, and respond to member needs around a complex tapestry of academic integrity issues pertaining to assessment design and security, online/remote teaching and learning practices, e-proctoring services, student demographics, internet connectivity and accessibility, contract cheating, file sharing websites, and student mental health and welfare. ACIFA takes official positions on issues that affect all members and significantly impact their "economic and professional well-being." Within the labour and welfare context, ACIFA undertook a series of initiatives to advocate for its members, including (but not limited to):

- educating members about contract cheating (Clarke & Lancaster, 2006), a term that had not made its way into the lexicon or wider awareness of faculty in Alberta's colleges and polytechnics prior to the pandemic,
- collaborating with the Alberta Council on Academic Integrity (ACAI) to lobby the provincial government to take action against contract cheating companies' exploitation and blackmailing of students,
- supporting instructors facing dramatic shifts in the composition of domestic and international students,
- providing resources about generative artificial intelligence (genAI) and the implications for valid assessment, and
- discussing strategies for supporting instructor welfare and reducing burnout.

In March 2023, "…the first position paper by a faculty association anywhere in the world on academic integrity" (Dr. Sarah Elaine Eaton, personal communication, 2023, March 17) was shared with ACIFA members and the wider community of academic integrity scholars and practitioners. ACIFA recommended 12 actionable steps to reduce the workload burden for instructors (many of which were described above). Although the position paper seeks to address a labour issue that affects its members in Alberta specifically, its recommendations for reducing the workload burden associated with academic integrity apply nationally and even globally to the working conditions of instructors in colleges and polytechnics grappling with increased academic misconduct cases.

Fostering a culture of academic integrity is a crucial yet demanding task. When instructors lack easy access to local faculty associations, advocacy groups, or academic integrity networks, the burden can become particularly onerous and can exacerbate feelings of isolation. However, several strategies can help instructors find support and share the workload. Here are some actionable steps instructors can take, and these are detailed in Stoesz et al. (2020) and Eaton et al. (2023b):

1. **Reach out and become involved in provincial, national, and international academic integrity groups. For example:**

   - **Alberta Council on Academic Integrity** (ACAI) https://albertaacademicintegrity.wordpress.com/
   - **British Columbia Academic Integrity Network** (BC AIN) https://bcain.trubox.ca/
   - **Manitoba Academic Integrity Network** (MAIN) https://mbacademicintegrit.wixsite.com/main
   - **Academic Integrity Council of Ontario** (AICO) https://sites.google.com/view/ai-council-ontario/home
   - **International Center for Academic Integrity** (ICAI)

     A global network that provides resources, research, and community support for academic integrity.
     https://academicintegrity.org/

   - **European Network for Academic Integrity** (ENAI)

An international organization that offers resources and networking opportunities to promote academic integrity in Europe and beyond.
https://www.academicintegrity.eu/wp/

These groups can also serve as exemplars for initiating local academic integrity networks in other regions. Establishing a local network can provide a platform for instructors to discuss challenges, share resources, and collaborate on initiatives to uphold academic integrity.

2. Start an Integrity-Hour style CoP as described by Eaton et al. (2021). A guide to developing and facilitating a community for academic integrity provides further details for getting started (Eaton, 2020).
3. **Organize an in-person, virtual, or hybrid event where academic integrity experts speak and answer audience questions (e.g.,** MAIN Speaker Series https://pupp.uqo.ca/en/event/main-speaker-series-2023-24/session1/**).** Hosting events featuring experts can enhance understanding and provide practical advice on managing academic misconduct.
4. **Seek out ready-to-use resources and advice from organizations devoted to promoting academic integrity:**
   - **TEQSA:** The Tertiary Education Quality and Standards Agency offers comprehensive guides and resources for protecting academic integrity. For more information: https://www.teqsa.gov.au/guides-resources/protecting-academic-integrity
   - **ENAI:** The European Network for Academic Integrity publishes valuable research and articles, such as: https://edintegrity.biomedcentral.com/articles/10.1007/s40979-023-00133-4

Finally, it is essential to remember that while fostering a culture of academic integrity is important, recognizing individual boundaries and limitations is crucial for personal well-being and preventing burnout. Balancing professional responsibilities with self-care ensures that instructors can continue to support academic integrity without compromising their mental and physical health.

Conclusion

Provincial and national advocacy efforts taking place in this third space of academic integrity work are productive and mitigate the ongoing and evolving challenges associated with fostering cultures of academic integrity and managing the increased physical and emotional labour associated with dealing with cases of academic misconduct. Advocacy efforts that seek to address structural workload issues in collective agreements and to shift the workload burden from individual instructors to a more distributed and resourced community of practitioners at the institutional, provincial, national, and even global levels will help to alleviate the inequitable workload burden. Shared community work yields dividends for individual sessional instructors, in particular, who shoulder the workload burden disproportionately. Shared community work is even more important with the ongoing impacts of genAI and other emerging technologies on teaching, learning, and academic integrity. Shared responsibilities and commitments support more equitable and ethical workload distribution among colleagues at colleges and polytechnics.

References

ACIFA. (2023). *ACIFA's position on academic integrity*. https://www.acifa.ca/_files/ugd/16fcf0_c98c6fb9f9bc4e4880a9f875d754eb4e.pdf

Ahuna, K. H., Frankovitch, L., & Murphy, G. (2023). Claiming space for honest work: Academic integrity as third space labor. *Workplace: A Journal for Academic Labor, 34*, 75–87. https://ices.library.ubc.ca/index.php/workplace/article/view/186840.

Clarke, R., & Lancaster, T. (2006). Eliminating the successor to plagiarism? Identifying the usage of contract cheating sites. In *Proceedings of 2nd plagiarism: Prevention, practice and policy conference 2006*. JISC Plagiarism Advisory Service.

Crossman, K. (2021). Education as a financial transaction: Contract employment and contract cheating. In S. E. Eaton & J. Christensen Hughes (Eds.), *Academic integrity in Canada: An enduring and essential challenge* (pp. 217–230). Springer. https://doi.org/10.1007/978-3-030-83255-1_11

Crossman, K. (2019). Is this in my contract? How part-time contract faculty face barriers to reporting academic integrity breaches. *Canadian Perspectives on Academic Integrity, 2*(1), 32–39. https://doi.org/10.11575/cpai.v2i1.68934

Dawson, P. (2021). *Defending assessment security in a digital world: Preventing e-cheating and supporting academic integrity in higher education*. Routledge.

Eaton, S. E. (2020). *Integrity hour: A guide to developing and facilitating an online community of practice for academic integrity*. University of Calgary. https://prism.ucalgary.ca/server/api/core/bitstreams/e646191b-4540-4a20-b88c-9a38ce431814/content.

Eaton, S. E., Stoesz, B. M., Crossman, K., Garwood, K., & McKenzie, A. (2023a). Faculty perspectives of academic integrity during COVID-19: A mixed methods study of four Canadian universities. *Canadian Journal of Higher Education, 52*(3), 42–58. https://journals.sfu.ca/cjhe/index.php/cjhe/article/view/189783.

Eaton, S. E., Stoesz, B. M., & McKenzie, A. (2023b). Academic integrity leadership and community building in Canadian higher education. In S. E. Eaton (Ed.), *Handbook of academic integrity* (pp. 1–21). Springer. https://doi.org/10.1007/978-981-287-079-7_171-1

Eaton, S. E., Moya Figueroa, B., Butterworth, C., Feledichuk, D., Leslie, K., Lothian, J., Murray, J., Soodeen, C., & Wheatley, B. (2021). Reflections on the first year of integrity hour: An online community of practice for academic integrity. *Canadian Perspectives on Academic Integrity, 4*(1), 4–13. https://doi.org/10.11575/cpai.v4i1.72676

Government of Alberta & The Alberta Union of Provincial Employees. (2021). Article 5 management recognition. *Collective agreement*. https://collective-agreement.alberta.ca/Pages/MyHR/Collective-agreement/Article-5-Management-Recognition.aspx.

Hamilton, M., & Wolsky, K. (2022). The barriers to faculty reporting incidences of academic misconduct at community colleges. In S. E. Eaton & J. Christensen Hughes (Eds.), *Academic integrity in Canada: An enduring and essential challenge* (pp. 467–485). Springer. https://doi.org/10.1007/978-3-030-83255-1_24

Labour Relations Code, R.S.A 2000, c L-1. https://open.alberta.ca/publications/l01

Murdoch, K. (2024, April 2). The nail that sticks up. *Guerilla Warfare*. https://www.guerillawarfare.net/post/the-nail-that-sticks-up.

Openo, J. (2024). Academic integrity and the affordances and limitations of authentic assessment. In S. E. Eaton (Ed.), *Second handbook of academic integrity* (pp. 217–231). Springer. https://doi.org/10.1007/978-3-031-54144-5_90

Openo, J. A., & Robinson, R. (2021). The emotional labour of academic integrity: How does it feel? *Canadian Perspectives on Academic Integrity, 4*(1), 20–39. https://doi.org/10.11575/cpai.v4i1.71350

Post-Secondary Learning Act, S.A. 2003, c P-19.5. https://open.alberta.ca/publications/p19p5

Stoesz, B. M., Seeland, J., MacLeod, P., Vogt, L., Markovics, L., Denham, T., Gervais, L., & Usick, B. (2020). Creating a collaborative network to promote cultures of academic integrity in Manitoba's post-secondary institutions. *Canadian Perspectives on Academic Integrity, 3*(1), 39–48. https://doi.org/10.11575/cpai.v3i1.69763

Vogt, L., & Eaton, S. E. (2022). Make it someone's job: Documenting the growth of the academic integrity profession through a collection of position postings. *Canadian Perspectives on Academic Integrity, 5*(1), 21–27. https://doi.org/10.11575/cpai.v5i1.74195

Brooklin Schneider is an educational developer and instructor teaching academic and technical writing at NorQuest College in Edmonton, Alberta. Brooklin completed a master's degree at the University of Waterloo. Brooklin is an active faculty association member and union advocate, working with local and provincial associations in Alberta to address concerns regarding academic integrity, workload, and other professional affairs. Brooklin is the author of the ACIFA (Alberta Colleges, Institutes, and Faculties Associations) Position on Academic Integrity. She is also a member of the Alberta Council on Academic Integrity Contract Cheating Working Group.

Chapter 10
From Mega to Micro: Collaborative Strategies for Enhancing Academic Integrity in a College Nursing Program

Josh Seeland, Lynn Cliplef, Mitchell Hengen, Kirk Joyce, and Kealy Murray

Abstract The 4M framework – comprising mega, macro, meso, and micro organizational levels - serves as this chapter's foundation and lens for an interdisciplinary and interdepartmental examination of the holistic approach to academic integrity within the Assiniboine College School of Nursing in Manitoba, Canada. The authors delve into each level, shedding light on their significance in fostering ethical congruence, enhancing college administration, and upholding professional standards. First, the college-run nursing program is examined at the mega level to identify features in the professional ethics of the sector and accrediting bodies. The focus then narrows to the institutional macro level, where the authors emphasize the importance of policy and procedure, which involves all stakeholders. Moving down the framework to the meso level, the authors explore the impact of interdepartmental collaboration in nurturing academic integrity, as well as examining the impacts of technology on teaching practices. Finally, at the micro level, the authors highlight the crucial role of individual faculty members, students, and staff in promoting academic integrity, with particular attention paid to challenges faced by nursing instructors. By instilling a sense of ethical responsibility and modelling integrity in daily actions, the authors illustrate how concerted and collaborative efforts can foster a culture of academic integrity. This example provides a model for how the 4M framework can be used in vocational training programs to develop a comprehensive approach to academic integrity that extends beyond the institution and into industry.

Keywords 4M framework · Institutional culture · College nursing · Professional standards · LPN training · Professional development · Academic integrity

J. Seeland (✉) · L. Cliplef · M. Hengen · K. Joyce
Assiniboine College, Brandon, MB, Canada
e-mail: seelandjl@assiniboine.net; cliplefl@assiniboine.net; hengenm@assiniboine.net; joycek@assiniboine.net

K. Murray
Assiniboine College, Portage la Prairie, MB, Canada
e-mail: murrayk4@assiniboine.net

L. Vogt, B. M. Stoesz (eds.), *Academic Integrity in Vocational and Polytechnic Education*, Ethics and Integrity in Educational Contexts 8,
https://doi.org/10.1007/978-3-031-90091-4_10

Canada is a country where education is decentralized; there is no federal ministry of education, nor is there a national quality assurance organization. Despite these challenges, academic integrity in Canadian higher education has been described as being firmly rooted in nationwide community development and strengths-based leadership (Eaton et al., 2024). At Assiniboine College in Brandon, Manitoba, Canada (hereafter referred to as Assiniboine), the School of Nursing has benefitted from this strengths-based leadership, which is rooted in a community-focused perspective on academic integrity. This chapter delves into these aspects through the foundational lens of the 4M Framework, dissecting the mega, macro, meso, and micro levels to highlight their importance in fostering ethical congruence, enhancing college administration, and upholding professional nursing standards. Educators in other vocational training programs can draw from this example by applying the same 4M Framework within their own contexts to develop a holistic approach to academic integrity that prepares students for the ethical professional responsibilities of their future careers.

The 4M Framework

Within the scholarship of teaching and learning, Bronfenbrenner's (1977) dynamic systems theory has been adapted as the 4M Framework and applied to the context of academic integrity in higher education by Eaton (2021) and later expanded upon by Kenny and Eaton (2022). Applying the 4M Framework to academic integrity enables a multi-lens organizational examination from the mega, macro, meso, and micro levels and is utilized in this chapter's discussion of academic integrity in Assiniboine's nursing program (Fig. 10.1).

The *mega* level occupies the highest and broadest level of the 4M Framework. In the context of Assiniboine, this level would include the surrounding community and higher education as a sector. Many individual stakeholders may, for one reason or another, feel detached from being able to enact change or action at this stratum.

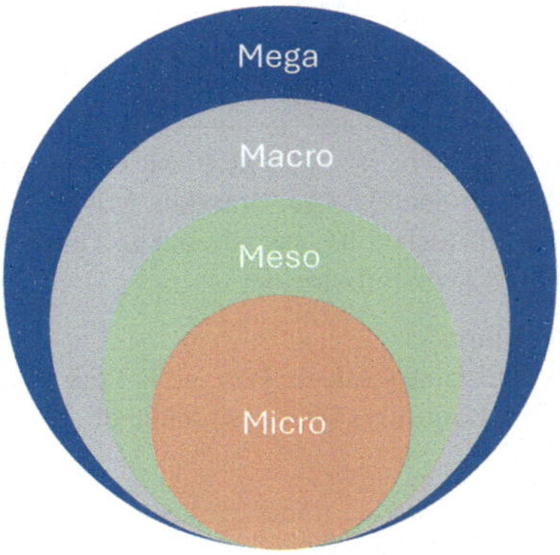

Fig. 10.1 The 4M framework

Kenny and Eaton (2022) specify technological, political, and economic factors as among those which wield expansive influence here. Pell and Amigud (2022) highlight that academic misconduct in a field such as nursing would be a concern at this level due to the social safety risk of the work being performed. Wideman (2010) points out that academic misconduct does not begin in post-secondary education; however, it is often the post-secondary educators who must address the issue and impart an understanding of how it is linked to the ethics of a professional. Ethics and values associated with professions, such as nursing, must be learned, and the link between academic misconduct and unethical practice is often left to individual instructors at the micro level to convey.

The *macro* level comprises the institution, its formally recognized structures, and some of the work done at senior leadership positions. In academic integrity, the macro level consists of policies, standards, and procedures that enable change and knowledge flow (Eaton, 2021; Kenny & Eaton, 2022). Sometimes, individuals feel detached from and powerless to participate at the macro level. When initiatives are driven from the meso and/or micro levels, senior leaders can avoid a "top-down" culture and work as "social architects" (Hannah & Lester, 2009) for systems within their organizations. Using what Bealle (2017) describes as a "holistic academic integrity strategy" (p. 146) empowers the individual to transcend through the framework levels.

The *meso* level at Assiniboine encompasses both formal and informal sub-organizations, with the former including the School of Nursing and the Learning Commons. Hannah and Lester (2009) characterize informal sub-organizations as "knowledge networks" which could be tasked to key positions that help to build knowledge bridges across formal organizational units and structures using shared leadership and semi-autonomous principles. Kenny and Eaton (2022) propose cross-unit working groups and committees as examples at this level. They also point out the "meso-level gap" (p. 580), which refers to the lack of focus on the potential for change in this level of departments, working groups, and faculties. The existence of invisible yet ubiquitous microcultures can either help or hinder actions at the meso level.

The *micro* level is one of the individual stakeholders. At Assiniboine, this would include students, instructors, administrators, and support staff. The chapter authors would occupy this level as individuals and in their respective roles. According to Hannah and Lester (2009), this level is the start and end of the learning process in organizations such as colleges since knowledge is created, disseminated, gathered, and codified in processes by the people within them. They further state that organizational leaders should promote "developmental readiness" in individuals through goal orientation, learning efficacy, and metacognitive ability. In a college, this may be a more frequent consideration for student stakeholders if they are primarily defined as the learners and staff as teachers, creating separation between the roles rather than focusing on the interconnected relationships between these two groups. Staff are also able to learn continuously in their area of expertise, as educators, and in numerous other ways through both formal and informal means. Kenny and Eaton (2022) further help dispel this possibility by suggesting that research, teaching, assessment, learning, and academic practices could be avenues for individuals to implement and model academic integrity, which presents many opportunities for staff. Pell and Amigud (2022) point out that senior leaders, as individuals, can enable institutional change by committing resources available within their positions.

Academic Integrity at the Mega Level

The Sector of Higher Education

Since higher education is decentralized in Canada, many institutions are left to create their own policies and procedures around quality assurance and academic integrity, and each will have differing approaches. Institutions and their members have multiple competing priorities and may not focus on examining academic integrity in their context (Bealle, 2017). Pell and Amigud (2022) describe faculty as feeling that academic integrity is being eroded, with institutions operating more like businesses rather than prioritizing learning. Further, Pell and Amigud also found that study respondents prioritized teaching and research over academic misconduct and related administrative activities. Thus, it is important to recognize that academic integrity is not always a top priority within academic institutions and that coordinated and sustained efforts may be required to maintain it.

Within the higher education sector, colleges often have less research focus than universities, as their priorities often focus on educating students for professional careers. This includes training students in professional practice and ethical behaviour. "The acculturation of students to the values of academic integrity supports their understanding of the requisites for ethical behaviour required for safe, professional practice" (Miron et al., 2021, p. 2). Thus, educating nursing students on academic integrity supports the development of ethical professional practice, and this education extends to other professions. Miron et al. (2021) continue, stating that "educating students in ethical practice is not unique to nursing and, in fact, is expected of many other professions like engineering (Engineers Canada, 2013), pharmacy (National Association of Pharmacy Regulatory Authorities, 2009), and physiotherapy (Physiotherapy Education Accreditation Canada, 2012), to name a few" (p. 2). However, the strong ethical component in nursing is advantageous for alignment with academic integrity initiatives. Examining the nursing profession, its ethics, and the role of regulatory bodies demonstrates that learning with integrity is not optional in nursing programs, which may combat fears regarding the increasingly transactional nature of education.

The Nursing Profession and Its Ethics

Academic integrity in nursing is commonly linked directly to the ethics of the nursing profession (Lynch et al., 2022; Milton, 2015; Miron, 2016, 2019); therefore, academic integrity may be viewed as the first building block in developing a professional moral compass while future professionals are still students (Lynch et al., 2022). Lynch et al. further state that academic integrity is essential in developing the critical thinking skills necessary for patient safety and honesty in one's professional practice. Nurse academics are viewed as the gatekeepers to the profession (Lynch et al., 2022), bearing the responsibility of preparing nursing students for practice in a way that aligns with nursing Standards of Practice - specific to each provincial regulatory body in Canada - which are rooted in nursing ethics (Emmanuel et al., 2021). Miron (2019) emphasizes the importance of instilling academic integrity values (honesty, trust, fairness, and respect) in nursing students, as these values are connected to professional behaviour in a clinical setting and contribute to ethical practice.

An example of how nurses in academic education can steer the moral compass of future nursing professionals is by guiding students to understand the opportunities and limitations of generative artificial intelligence (genAI), specifically text-generating tools that can respond to questions and formulate cohesive written responses. Nursing scholars are concerned that an overreliance on ChatGPT and other similar tools will impact the attributes that define the nursing profession, such as a culture of inquiry, leading to reduced critical thinking and, thus, potential reduction in developing the important wisdom that contributes to nursing intuition and ethical judgment (Abdulai & Hung, 2023). Inevitably, the ability to use genAI tools in a professional or scholarly way will become essential for nursing professionals and nurse scholars (Archibald & Clark, 2023); however, students and professionals must be aware of the potential for bias in genAI output. From a nursing educator lens, utilizing these biases as a teaching tool provides the opportunity to help students understand important ethics linked to the profession (Amsberry, 2022) and may assist in circumventing concerns related to overreliance on genAI by demonstrating its limitations. Additionally, nurses and other medical professionals play an important role in helping their patients determine the accuracy of generated output if they choose to use these tools to seek healthcare advice (Archibald & Clark, 2023), adding a new layer to the importance of nursing ethics, which must be addressed early in education, ideally before students begin practicum placements.

A unique finding in the literature regarding academic integrity and the nursing profession relates to the concept of caring. In Wideman's study (2019), nursing students showed how students' caring for one another and wanting the best for the group over themselves as individuals justified academic misconduct. Miron (2016) also expressed concerns about this topic, "...students may align departures from academic integrity within a caring reference" (p. 38). Miron highlighted that the year of the program a student is in may contribute to their understanding of professional ethics and their link to academic integrity. "Efforts aimed at promoting academic integrity and acculturating students to behavioural norms around the values associated with academic integrity may also establish core values that are consistent with nursing practice and essential to safe, ethical care" (Miron, 2019, p. 269).

The Role of Accrediting Bodies

Accrediting bodies can play a significant role in boosting academic integrity initiatives within a post-secondary institution, as standards of practice are often intertwined with academic integrity goals. The College of Licensed Practical Nurses of Manitoba (CLPNM) is the governing body of practicing Licensed Practical Nurses in Manitoba and ensures the Practical Nurse programs provide a curriculum that enables graduates to meet its Scope of Practice, including the Standards of Practice and Conduct and Code of Ethics. The CLPNM regulatory body has the legislated duty to evaluate practical nursing education programs against its Standards of Education to ensure:

- Practical Nurse programs meet the legislated regulatory standards for practical nurse education and practice
- Graduates acquire current entry-level competencies of practical nurses in Manitoba
- Graduates achieve the educational foundation to perform their duties at the current full scope of practice in Manitoba
- Graduates possess the judgement, behaviours, ethics, and attitudes required of a professional practical nurse in Manitoba
- Practical Nurse programs provide appropriate and adequate resources to support learning
- The educational institution is engaged in continuous development and improvement of the program (CLPNM, 2022a).

The CLPNM Education Approval Committee evaluation process reviews the program, its delivery, and its relationship with the hosting educational institution against the CLPNM Standards of Education for Practical Nursing Programs in Manitoba. There are explicit criteria within the categories of education standards by which practical nursing programs are measured to obtain program approval for nursing education delivery. These standards ensure that practical nursing programs in Manitoba provide high-quality education and training that prepares graduates to practice safely and effectively as licensed practical nurses.

As a result, ensuring institutional academic integrity is crucial for obtaining CLPNM approval to deliver practical nursing education. Clear policies and guidelines on academic integrity maintain the standards of honesty, transparency, and accountability in the program's development and delivery. Demonstrating a commitment to academic integrity signals to the CLPNM and healthcare stakeholders that the curriculum aligns with industry standards and regulatory requirements. This ensures public trust and confidence by producing competent and ethical LPNs dedicated to providing safe and competent client care.

Graduates of a CLPNM-approved Practical Nurse program are required to write the Canadian Practical Nurse Registration Exam (CPNRE) to obtain registration as an active practicing nurse in Manitoba. Applicants must hold a current Graduate Practical Nurse (GPN) registration with the CLPNM (CLPNM, 2022b) to be eligible to sit the exam. The registration exam ensures that potential Licensed Practical Nurse registrants have the necessary education and training to meet the CPNRE evaluation standards. The CLPNM reviews each GPN examination attempt for irregularities and misconduct, which, if found, may result in referral to the CLPNM Investigation Committee or, potentially, the Discipline Committee. The CLPNM notes that this process ensures public safety (CLPNM, 2022b). This aligns with Yardstick, the developer and issuer of the CPNRE, which states on its website, "The purpose of the exam is to protect the public by ensuring that the entry-level practical nurse possesses the competencies required to practice safely and effectively" (Yardstick, 2024). Meeting licensing exam standards validates the proficiency of graduate LPNs and assures employers and the public that these nurses have the necessary competencies to provide quality healthcare services. Given the high-stakes nature of the licensing exam and the protection of the public, the educational institution must uphold academic integrity throughout the nursing program.

The relationship between the accrediting body, the institution, and the student is therefore strengthened through academic integrity initiatives that cultivate a culture of honesty, responsibility, and ethical conduct among future healthcare professionals, translating to competent and safe client care. Upholding academic integrity in a college nursing program also contributes to long-term societal benefits by developing necessary competencies and skills for the nursing profession and fostering ethical decision-making (Guerrero-Dib et al., 2020). McClung and Gaberson (2021) state, "Not only may nursing students' dishonest behaviour build potential patterns of engaging in dishonest actions later in their careers, but also their dishonest behaviours may leave gaps in knowledge and understanding" (p. 112). This supports the notion that students who engage in academic misconduct during their training may not only be unsuccessful in the licensure exam but may also carry unsafe behaviours into nursing practice. This approach builds students' capacity to meet licensing exam standards for Licensed Practical Nurses in Manitoba and the nursing profession's ethical and professional standards, as well as safeguarding client safety and care.

Academic Integrity at the Macro Level

Institutional Academic Integrity Policy

Policies and procedures represent the institution's approach to addressing fundamental issues such as academic integrity at the macro level. These policies also serve as guiding documents that help shape an institution's overall culture and practices and serve as practical tools for governance and accountability. As Abbott and Nininger (2020) contextualize, "The academic integrity policies of nursing programs are shaped by systemic features of host institutions" (p. 270). Through continuous evaluation and refinement, these policies evolve to meet the changing needs and challenges of the academic community. It should be noted that policy functions best in conjunction with education for all stakeholders and, in the context of academic integrity, other proactive and preventative methods (Kier & Ives, 2022). Wideman (2009) describes this as a "shift from the traits of individual students to the situations which provide the students opportunities for cheating" (p.159). Contributing to policy development allows individuals to see themselves reflected in institutional documents and take ownership of the guidelines that govern their daily practice.

Examining academic integrity policy and procedure can be used to categorize an institution's approach to academic integrity (Bertram Gallant, 2011). At Assiniboine, members of the Learning Commons, which is part of the meso level, drew from research on academic integrity policy (Bretag & Mahmud, 2016; Morris & Carroll, 2011, 2016) to examine Assiniboine's Policy A25 and identify a need for revision based on its rule compliance approach. Significant areas of focus in the revision included replacing negative, punitive, and criminological language with neutral and, where possible, supportive terminology. Examples include removing "honesty" and "dishonesty" in favour of "misconduct" and removing the word "guilty." Additionally, it was important to move away from an assumed burden of proof model to one of a balance of probability, in which decision-makers are supported in determining if it is more likely than not that academic misconduct has taken place. Modernized categories and examples of academic misconduct were added to the policy. This included updating the term "essay mill" to "contract cheating" and providing clarification and nuance to this pervasive form of academic misconduct. Other policy changes included consideration of rephrasing tools, academic file-sharing sites, and the misuse of genAI tools. A final feature of the revised Policy A25 is that it has been separated from the new Academic Misconduct Procedure Manual. The latter document provides instructors and administrators from the School of Nursing and all other academic departments at Assiniboine with clear steps and forms to make the "reporting of student misconduct easier" (Dannhoferova et al., 2022, p.18). Assiniboine found that changing the language used at the macro level had an impact on how academic integrity is perceived and discussed at other 4M levels. Further description of the cultural change toward a positive academic integrity environment is described by Munn et al. (2025).

Academic Integrity at the Meso Level

The Learning Commons

A unique aspect of Assiniboine is that the library, teaching and learning centre, and student academic support departments are combined into the *Learning Commons*, forming a central unit to coordinate and lead activities at the meso level. This integration allows the departments to work together as a team, providing interdisciplinary and collaborative support for teachers, staff, and students and supporting academic integrity throughout all stages of assessment, including design and development, implementation, evaluation of student work, and reviewing and revising assessments. By taking a collaborative role in promoting academic integrity with different stakeholder groups, the Learning Commons bridges the gap between individual actions at the micro level and institutional policies at the macro level and promotes the importance of all institutional members in advancing academic integrity. The Learning Commons has also been instrumental to the Academic Integrity Advisory Committee and the redevelopment of academic integrity policy and procedure.

It is important to understand the multi-faceted role of the Learning Commons and its impact on the institution. During the assessment design and development stages, the teaching and learning centre, known as the Centre for Learning and Innovation or CLI, works with the curriculum developers and instructors to create assignments that help prevent academic misconduct from occurring and optimize the genuine achievement of learning outcomes. After the assessments are designed, CLI supports instructors in educating students on academic integrity. The Assiniboine Library can provide direct student support through in-class or online workshops and instruction or face-to-face support within the library. This follows the assertion by Miron et al. (2021) that library staff in Canada are the group most commonly tasked with providing academic integrity instruction for students besides their instructors.

During the evaluation of student work, the Assiniboine Library supports instructors and academic administration by identifying potential academic misconduct through bibliographic analysis and discovery interviews. Bibliographic analysis examines the nature and application of sources cited in submitted student work. Discovery interviews are discussions between college staff and students to learn more about the students' process when there is a concern about academic misconduct (Eaton, 2019). This is consistent with Dannhoferova et al. (2022), who found that libraries can work to support the investigation of possible cases of academic misconduct. Following the semester, CLI helps identify assessments with reported academic misconduct issues and works with instructors to update the assignments to help prevent and reduce future academic misconduct.

The Learning Commons also provides professional development for faculty and staff at the college regarding academic integrity, which Miron (2016) specifies as a crucial part of cultures of academic integrity. CLI facilitates a voluntary program aimed at new faculty called the Teaching Essentials Program (TEP). This course introduces the college's positive, educative approach to academic integrity, emphasizing that it is the responsibility of everyone in the college community. The TEP course is facilitated via Zoom, allowing instructors at all campuses and rotating sites to participate. Another training opportunity offered to faculty at the college is Course Design Blueprinting (CDB), in which participants are walked through the process of designing a course using a blueprint. Academic integrity is included as an integral part of course and assessment design. During the institutional onboarding process organized by Assiniboine's Human Resources department, all new staff are introduced to the college culture of academic integrity through the Learning Commons' presentation. Patterson et al. (2003) suggests that academic integrity instruction be linked to professional and ethical issues as an institutional priority, which Learning Commons staff seek out when delivering sessions for specific programs and academic areas.

Finally, the Learning Commons has an impact on the meso level by providing professional development opportunities for higher education professionals outside of Assiniboine, organizing events in collaboration with the Manitoba Academic Integrity Network (MAIN), including the annual meeting and a speaker series (see Seeland, 2021). The 7th AIIIM event in 2024 was hosted at one of Assiniboine's School of Nursing rural sites, bringing representatives from colleges and universities across Manitoba and Saskatchewan together. This site selection was intentional to better provide professional development and networking opportunities for staff working outside of major urban locations as well as those within them.

Assiniboine's Academic Integrity Advisory Committee

Committees can play a vital role in promoting institutional values and norms and establishing an avenue for collaboration within and across the institution at the meso level of the 4M framework. Striking committees mandated with academic integrity has been a recommendation in academic integrity literature for quite some time (Whitley & Keith-Spiegel, 2001). Through initiatives such as professional development opportunities, these committees serve as conduits for disseminating information and best practices related to academic integrity throughout the organization. Situating committees at the meso-level also helps prevent a top-down detachment among stakeholders.

Assiniboine's Academic Integrity Advisory Committee (AIAC) was created in late 2019 at the request of the Vice President, Academic. Its terms of reference are to inform Learning Commons' services, projects, and resources, identify opportunities to implement and evaluate academic integrity interventions and supports, and contribute to policy and procedure development. Membership is selected to facilitate broad institutional engagement of stakeholders from each academic school and campus, along with the Student Association and as many academic staff positions as possible. In this capacity, the AIAC manifests how Assiniboine has come to view "academic integrity as a multi-stakeholder responsibility" (Kenny & Eaton, 2022, p. 574) rather than a problem with dishonest students. The committee is also an example of Hannah and Lester's (2009) "social architect" approach to macro-level actions by senior leaders, enabling meso-level "knowledge networks" across formal organizational units. By embodying this approach, the AIAC effectively spreads information about academic integrity throughout the college organization – from professional development opportunities to the latest research.

The AIAC has directly benefited Assiniboine's School of Nursing by creating consistency across multiple student intakes at urban campuses and rural and rotating sites. Many locations are physically separated from the amenities and immediate support services of the larger campuses, leading to past concerns about inconsistent academic integrity approaches. When instructors and administrators from multiple sites began participating in the AIAC, they were able to enhance their ability to uphold standards and support students in maintaining ethical conduct by staying abreast of the latest developments and best practices in academic integrity. This facilitates the modelling of academic integrity for students at each location, and co-workers benefit from on-site support and guidance. Additionally, having representation from all nursing stakeholders ensures that the unique challenges and dynamics of rural and rotating nursing sites are considered in the committee's initiatives and strategies. This inclusivity helps bridge potential gaps in support and resources and ensures that stakeholders receive consistent and equitable assistance regardless of their location. Ultimately, the presence of nursing staff on the AIAC contributes to fostering a culture of integrity that is responsive to the diverse needs of stakeholders across all academic settings within the institution.

Teaching Practices

Professional training programs often rely on longstanding approaches to teaching and assessing, and examining these more closely in light of new teaching methods or new avenues for academic misconduct supports improvements at the meso level. In nursing education, standardized patient activities are vital elements. In this learning activity, laypeople portray patients presenting concerns and nursing students practice various aspects of nursing care. In recent years, virtual simulations (vSIMS) patients have been offered to roleplay interactions between nursing students and patients in cases where students cannot find a suitable candidate to interview for standardized patient assignments. The virtual simulation programs allow students to have a tailored experience based on their choices throughout the program. Most programs allow the instructor to see each student's progress in some way; however, the details will be based on the program. It can be difficult and time-consuming for an instructor to review the detailed overview of the student's progress throughout the simulation. Further to this, some programs chose to utilize a list of questions which has a standard set of answers. This is problematic for the potential replication of answers. Moreover, these programs are sold throughout North America, and assignments can be reproduced or shared through assignment-sharing websites. Thus, using virtual simulation standardized patients for nursing assignments should be carefully and conscientiously thought through. When departments identify potential avenues for academic misconduct, such as this, there is a further collective responsibility to meet the changing landscape and continue to uphold ethical standards.

A solution to reduce this problem may be to create an assignment that focuses on personal reflection about the *process* of going through the simulation. The skill of personal reflection is often utilized as a source to enhance critical thinking in nursing practice (Bulman et al., 2014). Practicing critical reflection is an essential skill in learning to become a nurse (Smith, 2011). When used as an assessment tool, reflective practice allows students to delve more deeply into values and ethical principles related to nursing practice (Siles-González & Solano-Ruiz, 2016). Thus, it may be valuable for nursing programs to consider providing a higher weighted grade to these essential personal reflection activities to help foster critical thinking and professionalism (Smith, 2011) while also upholding both the ethics and standards of the profession and contributing in a positive way to academic integrity standards.

Academic Integrity at the Micro Level

The Curriculum Coordinator Position

A crucial aspect of promoting academic integrity in the Assiniboine School of Nursing lies in the role of the Curriculum Coordinator. This position helps address academic integrity at the micro level and bridges the meso level by unifying curriculum learning outcomes and assessments for the nursing program. The Curriculum Coordinator position is pivotal in promoting equity, diversity, inclusiveness, and cultural and psychological safety through assessment design, thus avoiding a "one size fits all approach" (Emmanuel et al., 2021, p. 50) that can contribute to academic misconduct. The role includes reviewing the assessment design of course content and acknowledging the strong correlation between academic dishonesty and assessment design (Emmanuel et al., 2021).

Quality assurance and quality improvement are also key aspects of this position, enhancing compliance with academic integrity as all issues with academic misconduct related to course content are reported to the Curriculum Coordinator for assignment design review. This role also bridges the meso level by liaising with external partners and the regulatory body to adapt and build curriculum at a micro level. The Curriculum Coordinator guides the planning of curriculum renewal and design with college staff, industry partners, and other educational institutions. Additional responsibilities include assisting with implementing and evaluating new initiatives, launching and evaluating new courses and programs, and developing and delivering professional development related to curriculum for faculty.

Nursing Instructors

The importance of nursing instructors upholding and adhering to academic integrity principles is often connected to the significance of upholding the Standards of Practice that align with the profession (Emmanuel et al., 2021). Miron (2022) emphasizes the opportunity afforded by work-integrated learning to "acculturate students to the values consistent with academic integrity and ethical, professional work" (p. 240), making nursing instructors, in particular, responsible for a unique part of bridging nursing students into practicing nurses. Many nursing students report relying on their instructors to teach them about academic integrity (Miron et al., 2021). "Role modelling by nurse academics is important for students' personal and professional development" (Emmanuel et al., 2021, p. 49). Wideman's study (2009) also pointed out that students will notice when academic integrity is not being modelled by instructors, such as when images used in presentations are not cited properly.

The School of Nursing has developed Policy SON-03 Professional Suitability, which compliments the requirements of the college-wide policy A25. Policy SON-03 makes a connection between learning and nursing practice to guide students and faculty, stating:

> ... Professional Nursing students must demonstrate that they are worthy of public trust, are engaged, caring, compassionate, and skilled professionals (within acquired competencies). Infractions to this policy, including issues of clinical competencies ... academic integrity, and student conduct, may be directed to the Professional Suitability committee for possible disciplinary action ... The outcomes/decisions taken as a result of a professional suitability hearing will be forwarded to the CLPNM for further review and possible suspension of the student license.

Faculty can use this complementary policy to help students understand the link between the ethics of the profession and academic integrity. However, nursing students also need guidance to differentiate caring for others as part of professional practice and helping classmates in an educational setting, as caring that becomes sharing can result in academic dishonesty (Miron, 2016; Wideman, 2009, 2011).

College instructors face various challenges regarding the implementation of academic integrity policies. The first challenge is that the teaching load is usually significantly higher than universities or even larger research colleges (Bealle, 2017), and thus, due to time constraints, instructors often want "someone to solve the academic issues for them, or at least provide advice" (Dannhoferova et al., 2022, p.15). Other challenges individual instructors may encounter are a lack of time to deal with academic misconduct (Pell & Amigud, 2022), feeling pressured to pass students, fear of student evaluation repercussions that impact personal yearly performance reviews, and convoluted reporting systems (Emmanuel et al., 2021).

Additionally, Canadian nursing shortages and the expansion of nursing programs intensify the demand for qualified instructors (Boamah et al., 2021). Penn et al. (2008) reported a rapidly aging nursing workforce, an increasing demand for care, and "significant salary differentials between academic and practice" (Penn et al., 2008, p. 2). Papazian et al. (2023) found that burnout among nurses since the COVID-19 pandemic is higher, adding another form of attrition. These factors lead to challenges, including finding qualified individuals, navigating limited lead times, and providing adequate professional development and mentorship. Instructors may be hired based on clinical experience and unique, individual skill sets associated with these clinical experiences.

Teaching allows industry experts to pass on their knowledge and enjoy stable work hours. Penn et al. (2008) report that some nursing instructors are motivated to teach by the desire to shape the next generation of nurses. However, transitioning from a practitioner to an educator involves significant adjustments and the acquisition of new skill sets. Mentors can assist with some adjustments by connecting new faculty to learning values within a department, including academic integrity, and ensuring that learning cultures are valued and upheld. Informal mentors can be found through participation in college committees and professional development events. This can assist faculty in establishing channels across the "Meso-level gap" (Kenny & Eaton, 2022, p. 580) for communication and collaboration in both structured and emergent ways. Reflecting on academic integrity initiatives across the 4M Framework highlights the importance of making academic integrity part of staff onboarding and orientation processes to help the micro level meet the macro level, where nursing instructors feel supported in upholding academic integrity policy regardless of experience or provincial location.

Conclusion

In conclusion, this chapter underscores the vital importance and challenges of adopting a comprehensive and collaborative approach to fostering academic integrity within a nursing program. By utilizing the 4M Framework, the authors illustrate how efforts at the mega, macro, meso, and micro levels collectively contribute to creating a robust culture of academic integrity that prepares students to meet their professional obligations with the highest ethical standards. Identifying the stakeholders and supports available within the institution and the profession, then mapping them with the 4M Framework, helps establish connections that foster a holistic approach to academic integrity. This holistic approach offers numerous benefits: staff and faculty are better supported in their practice, and graduates enter their profession skilled, knowledgeable, and ethically grounded, ready to face the challenges of their careers with integrity.

The practice of multi-disciplinary reflection, as our team experienced while preparing this chapter, creates opportunities for continuous improvement at each of the 4M levels. Staff and faculty from regulated professional training programs may find similar value in applying the 4M Framework as a lens for their collective academic integrity strategy. We hope this work will serve as a model for other multidisciplinary teams reflecting on academic integrity within their own institutions.

References

Abbott, M. R. B., & Nininger, J. (2020). Academic integrity in nursing education: Policy review. *Journal of Professional Nursing, 37*(2), 268–271. https://doi.org/10.1016/j.profnurs.2020.12.006

Abdulai, A. F., & Hung, L. (2023). Will ChatGPT undermine ethical values in nursing education, research, and practice? *Nursing Inquiry, 30*(3). https://doi.org/10.1111/nin.12556

Amsberry, S. (2022). Promoting academic integrity in nursing education: An integrative review. *Journal of Nursing Education, 61*(6), 303–307.

Archibald, M. M., & Clark, A. M. (2023). ChatGTP: What is it and how can nursing and health science education use it? *Journal of Advanced Nursing, 79*(10). https://doi.org/10.1111/jan.15643

Bealle, P. (2017). Community college academic integrity lessons that put research into practice. *Theory Into Practice, 56*(2), 144–115. https://doi.org/10.1080/00405841.2017.1283573

Bertram Gallant, T. (2011). *Building a culture of academic integrity* [White paper]. Magna Publications. https://www.depts.ttu.edu/tlpdc/Resources/Academic_Integrity/files/academicintegrity-magnawhitepaper.pdf.

Boamah, S. A., Callen, M., & Cruz, E. (2021). Nursing faculty shortage in Canada: A scoping review of contributing factors. *Nursing Outlook, 69*(4), 574–588.

Bretag, T., & Mahmud, S. (2016). A conceptual framework for implementing exemplary academic integrity policy in Australian higher education. In T. Bretag (Ed.), *Handbook of academic integrity*. Springer.

Bulman, C., Lathlean, J., & Gobbi, M. (2014). The process of teaching and learning about reflection: Research insights from professional nurse education. *Studies in Higher Education, 39*(7), 1219–1236.

College of Licensed Practical Nurses of Manitoba. (2022a). *Education program evaluation & approval.* Retrieved June 8, 2024, from https://www.clpnm.ca/for-employers-and-educators/educator-resources/education-program-approval/

College of Licensed Practical Nurses of Manitoba. (2022b). *Canadian Practical Nurses Registration Examination.* Retrieved June 8, 2024, from https://www.clpnm.ca/for-registrants/students-and-graduates/graduates/cpnre/

Dannhoferova, J., Foltýnek, T., Dlabolova, D., & Fishman, T. (2022). Cutting the vicious circle: Addressing the consistency in teachers' approaches to academic integrity breaches. *Journal on Efficiency and Responsibility in Education and Science, 15*(1), 10–22. https://doi.org/10.7160/eriesj.2022.150102

Eaton, S. E. (2019). *How to lead a discovery interview about contract cheating*. The University of Calgary, Taylor Institute for Teaching and Learning. https://taylorinstitute.ucalgary.ca/resources/how-to-lead-discovery-interview-contract-cheating

Eaton, S. E. (2021). *Plagiarism in higher education: Tackling tough topics in academic integrity.* Libraries Unlimited.

Eaton, S. E., Stoesz, B. M., & McKenzie, A. (2024). Academic integrity leadership and community building in Canadian higher education. In S. E. Eaton (Ed.), *Second handbook of academic integrity* (pp. 1847–1866). Springer Nature. https://doi.org/10.1007/978-3-031-54144-5_171

Emmanuel, E., Fielden, J., & Miller-Rosser, K. (2021). Why should we care about academic integrity in nursing students. *Journal of Nursing Education and Practice, 11*(5), 46–53.

Guerrero-Dib, J. G., Portales, L., & Heredia-Escorza, Y. (2020). Impact of academic integrity on workplace ethical behavior. *International Journal for Educational Integrity, 16*(2), 1–18. https://doi.org/10.1007/s40979-020-0051-3

Hannah, S. T., & Lester, P. B. (2009). A multilevel approach to building and leading learning organizations. *The Leadership Quarterly, 20*(1), 34–48. https://doi.org/10.1016/j.leaqua.2008.11.003

Kenny, N., & Eaton, S. E. (2022). Academic integrity through a SoTL lens and 4M framework: An institutional self-study. In S. E. Eaton & J. Christensen Hughes (Eds.), *Academic integrity in Canada: Ethics and integrity in educational contexts* (Vol. 1). Springer. https://doi.org/10.1007/978-3-030-83255-1_30

Kier, C. A., & Ives, C. (2022). Recommendations for a balanced approach to supporting academic integrity: Perspectives from a survey of students, faculty, and tutors. *International Journal for Educational Integrity, 18*(22). https://doi.org/10.1007/s40979-022-00116-x

Lynch, J., Glew, P., Salomonson, Y., & Ramjan, L. N. (2022). "Integrity is integrity. It doesn't matter the context": A qualitative exploratory study of academic integrity in an undergraduate nursing program. *Teaching and Learning in Nursing, 17*(4). https://doi.org/10.1016/j.teln.2022.06.013

McClung, E., & Gaberson, K. (2021). Academic dishonesty among nursing students. *Nurse Educator, 46*(2), 111–115.

Milton, C. L. (2015). Ethics and academic integrity. *Nursing Science Quarterly, 28*(1). https://doi.org/10.1177/0894318414558620

Miron, J. (2016). *Academic integrity and senior nursing undergraduate clinical practice.* [Doctoral dissertation. Queen's University]. https://qspace.library.queensu.ca/items/f55a08a4-b77b-47ee-b14e-3cf5ffc469f4.

Miron, J. (2019). Academic integrity in a student practice environment—an elicitation study. *The Journal of Educational Thought (JET), 52*(3), 253–274. https://www.jstor.org/stable/26873109

Miron, J. (2022). Academic integrity in work-integrated learning (WIL) settings. In S. E. Eaton & J. Christensen Hughes (Eds.), *Academic integrity in Canada. Ethics and integrity in educational contexts* (Vol. 1). Springer. https://doi.org/10.1007/978-3-030-83255-1_12

Miron, J., Eaton, S. E., McBreairty, L., & Baig, H. (2021). Academic integrity education across the Canadian higher education landscape. *Journal of Academic Ethics, 19*, 441–454. https://doi.org/10.1007/s10805-021-09412-6

Morris, E. J., & Carroll, J. (2011). JISC Policy works: Recommendations for reviewing policy to manage unacceptable academic practice in higher education. https://s3.eu-west-2.amazonaws.com/assets.creode.advancehe-document-manager/documents/hea/private/policy_works_0_1568037043.pdf.

Morris, E. J., & Carroll, J. (2016). Developing a sustainable holistic institutional approach: Dealing with realities "on the ground" when implementing academic integrity policy. In T. Bretag (Ed.), *Handbook of academic integrity*. Springer.

Munn, C., Rexe, D., & Seeland, J. (2025). Successful institutional change: Lessons on building a culture of academic integrity. In L. Vogt & B. M. Stoesz (Eds.), *Academic integrity in vocational and polytechnic education*. Springer.

Papazian, L., Hraiech, S., Loundou, A., Herridge, M. S., & Boyer, L. (2023). High-level burnout in physicians and nurses working in adult ICUs: A systematic review and meta-analysis. *Intensive Care Medicine, 49*(4), 387–400. https://doi.org/10.1007/s00134-023-07025-8

Patterson, B., Taylor, L., & Usick, B. (2003). The construction of plagiarism in a school of nursing. *Learning in Health and Social Care, 2*(3), 147–158.

Pell, D. J., & Amigud, A. (2022). The higher education dilemma: The views of faculty on integrity, organizational culture, and duty of fidelity. *Journal of Academic Ethics, 21*(4). https://doi.org/10.1007/s10805-022-09445-5

Penn, B. K., Wilson, L. D., & Rosseter, R. (2008). Transitioning from nursing practice to a teaching role. *The Online Journal of Issues in Nursing, 13*(3) https://ojin.nursingworld.org/table-of-contents/volume-13-2008/number-3-september-2008/nursing-practice-to-teaching-role/

Siles-González, J., & Solano-Ruiz, C. (2016). Self-assessment, reflection on practice and critical thinking in nursing students. *Nurse Education Today, 45*, 132–137.
Seeland, J. (2021). Spotlight: Manitoba Academic Integrity Network speaker series. *Integrity Matters*. https://academicintegrity.org/resources/blog/71-2021/september-2021/300-spotlight-manitoba-academic-integrity-network-speaker-series
Smith, E. (2011). Teaching critical reflection. *Teaching in Higher Education, 16*(2), 211–223. https://proxy.assiniboine.net:4132. https://doi.org/10.1080/13562517.2010.515022
Whitley, B. E., & Keith-Spiegel, P. (2001). Academic integrity as an institutional issue. *Ethics & Behavior, 11*(3), 325–342. https://doi.org/10.1207/S15327019EB1103_9
Wideman, M. (2009). *Caring, sharing, coping and control: Academic dishonesty and the nursing student*. [Doctoral dissertation: University of Toronto]. https://tspace.library.utoronto.ca/handle/1807/19139.
Wideman, M. (2010). *Academic dishonesty in a school of nursing ICETI 2010: The international conference on education*. Training and Informatics. http://www.iiis.org/CDs2010/CD2010IMC/ICETI_2010/PapersPdf/EB142ZD.pdf
Wideman, M. (2011). Caring or collusion? Academic dishonesty in a school of nursing. *Canadian Journal of Higher Education, 41*(2), 28–43. https://journals.sfu.ca/cjhe/index.php/cjhe/article/view/2298/2216
Yardstick. (2024). *The purpose of the CPNRE*. Retrieved May 6, 2024 from https://www.cpnre.ca/.

Josh Seeland is the Manager of Library Services at Assiniboine College, a founding member of the Manitoba Academic Integrity Network (MAIN), and Chair of Assiniboine's Academic Integrity Advisory Committee. His writing on academic integrity can be found in academic journals, and the books "Contract Cheating in Higher Education," and "Academic Integrity and the Role of the Academic Library: Institutional Examples and Promising Practices."

Lynn Cliplef is a Program Review Specialist for the Centre for Learning and Innovation at Assiniboine. She is interested in the intersection between the fields of teaching and learning in higher education and academic integrity. Her writing can be found in the International Journal of Mathematical Education in Science and Technology and Collective Essays on Learning and Teaching.

Mitchell Hengen is a Faculty Development Coach for the Centre for Learning and Innovation at Assiniboine. He is passionate about supporting faculty to find creative solutions around developing learning activities and assessments that support a positive culture of academic integrity to improve the learning experience.

Kirk Joyce is the Curriculum Coordinator for the School of Nursing at Assiniboine College. His experience working with learners and curriculum in various positions at the college ignited his passion for shaping innovative, learner-centered curriculum that upholds academic integrity and prepares future healthcare professionals.

Kealy Murray is an instructor in Assiniboine's School of Nursing and a member of its Academic Integrity Advisory Committee. Kealy is a Registered Nurse with a Master of Nursing degree.

Chapter 11
Strength of Character: Upholding Academic Integrity in Canadian Military Vocational Training

Catherine E. Déri, Caroline Kirouac, and Lucie Moussu

Abstract The Canadian Armed Forces employ 68,000 military members, categorized into two distinctive groups: the officers, who hold university degrees, and the technicians, who are qualified for one of 65 trades. Upon recruitment, these technicians complete basic military training and vocational training to perform common tasks and specialized roles. Then, they engage in professional development activities to assume increasing leadership responsibilities based on their potential for career progression. As such, the intermediate, advanced, and senior leadership programs offered by one of the three military colleges in Canada aim to develop knowledge and competencies for the technicians to attain the highest standards of the Profession of Arms. Among the topics covered in these programs is the Military Ethos with its six values that all military personnel are expected to exhibit on and off duty: loyalty, integrity, courage, excellence, inclusion, and accountability. This chapter will present a reflection on the value of integrity from an academic perspective, when military technicians take part in vocational training and professional development programs. We will first discuss how academic integrity is addressed by military institutions delivering vocational training and professional development programs. Next, we will compare the definitions of integrity developed by the Canadian Armed Forces and the International Center for Academic Integrity (ICAI). Finally, we will explore future considerations for integrating artificial intelligence (AI) and its impact on academic integrity in the Canadian Armed Forces education and training framework.

C. E. Déri (✉)
Université du Québec en Outaouais, Gatineau, QC, Canada
e-mail: catherine.deri@uqo.ca

C. Kirouac
Université du Québec à Rimouski, Lévis, QC, Canada
e-mail: caroline.kirouac@uqar.ca

L. Moussu
Royal Military College of Canada, Kingston, ON, Canada
e-mail: moussu@rmc.ca

L. Vogt, B. M. Stoesz (eds.), *Academic Integrity in Vocational and Polytechnic Education*, Ethics and Integrity in Educational Contexts 8,
https://doi.org/10.1007/978-3-031-90091-4_11

Keywords Academic integrity · Vocational education · Military training · Integrity values · Professional development · Military ethos · Canada

Introduction

The Canadian Armed Forces (CAF) provide vocational training to military technicians who specialize in 65 different trades, such as ammunition technician, aircraft maintenance, and sonar operator. These technicians are soldiers, sailors, and aviators who become skilled experts through specialized training and hands-on experience to partake in military operations in Canada or abroad. In addition to the three military colleges in Canada, there are about 115 training establishments that deliver vocational training to military members at the start of their careers after the obtention of a Basic Military Qualification (BMQ). During this initial training, all military members learn about the values associated with the Military Ethos: loyalty, integrity, courage, excellence, inclusion, and accountability (Canadian Defence Academy [CDA], 2022). It is expected that military technicians will embrace these values and adhere to their inherent principles, be it in their professional or personal lives. This means that military technicians should show integrity in their temporary roles as learners, even when there is little to no focus on academic integrity in some of the learning environments. This chapter presents a reflection on the value of integrity from an academic perspective when military technicians take part in vocational training and professional development programs throughout their careers. First, we explain how academic integrity is integrated into the CAF Professional Development Framework. Then, we discuss the commonalities and distinctions between the fundamental values of academic integrity and the Military Ethos. Finally, we explore the impact that generative artificial intelligence (GenAI) may have on academic integrity in military training, considering the most recent AI guidelines and strategies disseminated by the Government of Canada and the Department of National Defence (DND).

The CAF Professional Development Framework

In the CAF, military training is designed to produce well-rounded and highly skilled individuals, capable of serving effectively in a variety of operational environments while upholding the values and traditions of the Canadian military. To this effect, the organization has established a comprehensive framework with policies, instructions, guidelines, and specifications, along with a developmental structure to facilitate the successful development of military members from recruitment to retirement. As depicted in Fig. 11.1, the Canadian Armed Forces Professional Development System (CAFPDS) is based on the fundamental principles of a professional body of knowledge, war fighting skills, and leadership competencies. The policy framework

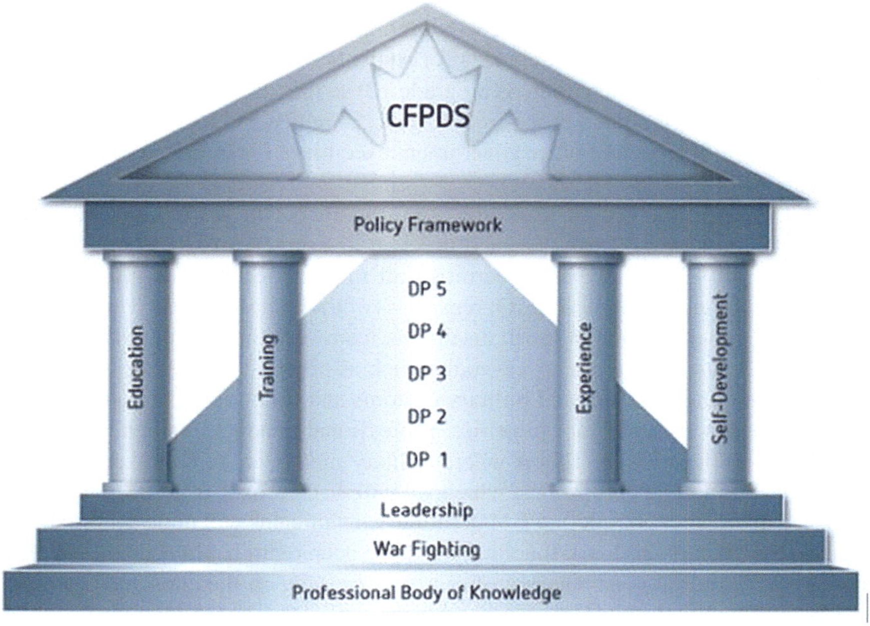

Fig. 11.1 The four pillars of the Canadian armed forces professional development system (CAFPDS) (This illustration of the CAF professional development system is being updated to reflect the description that is provided in this section of the chapter. The new illustration will be published by the Department of National Defence in 2025)

is then divided into four pillars that are designed to complement one another: education, training, employment experience, and self-development. At the centre of the framework are five developmental periods (DP) that military members will progress through during their careers.

The *Education* pillar is defined as "the provision of a base of knowledge and intellectual skills upon which information can be correctly interpreted and sound judgement exercised" (Government of Canada, to be published, section 4.2). This pillar includes both academic education and Professional Military Education (PME), which is akin to professional development. Regarding formal education, military technicians are expected to have completed grade 10 before being recruited. However, there is an increasing number of new recruits who have completed college or university programs (Bériault, 2021). When that is not the case, technicians are encouraged to pursue post-secondary education during their military career. This can be done at one of the three military colleges, namely the Royal Military College of Canada (RMCC) in Kingston, the Collège militaire royal (CMR) in Saint-Jean, or the Canadian Forces College (CFC) in Toronto. The RMCC is a military university offering undergraduate and graduate programs in Social Sciences and Humanities, Sciences, and Engineering. Unlike civilian universities in Canada, all

RMCC undergraduate degrees include a "common core" in Canadian History, Language and Culture, Political Science, International Relations, and Ethics. As for CMR, it provides college-level courses for students who completed high school in the Quebec province (where there is no grade 12) so that they can successfully transition into an undergraduate program at RMCC. Finally, CFC focuses on preparing senior military officers and other government executives for future strategic leadership responsibilities.

The *Training* pillar is defined as "the provision of specific skills, knowledge, and attitudes required to perform assigned tasks" (Government of Canada, to be published, section 4.2). After being recruited, all military technicians must undergo individual training to obtain a BMQ, regardless of their chosen trade. This demanding training program aims to instill discipline, teamwork, and basic military skills. To that effect, its curriculum covers topics such as physical fitness, dress and drill, weapons handling, first aid, and military customs and traditions. The foundation of BMQ training rests upon four priorities: professional conduct, resiliency, physical fitness, and military skills. This is where military members are first exposed to the Military Ethos and its core values, which we will discuss further in the next section of this chapter. After completing BMQ, military technicians will undergo vocational training based on the military function they have been recruited to accomplish. This trade-specific training can vary significantly, depending on the chosen occupation, and is designed to provide members with the skills and knowledge necessary to excel in their trade (Rounding et al., 2018). As previously mentioned, to provide the required training, the CAF operates a comprehensive network of about 115 training establishments located across Canada. This network facilitates, among other things, the delivery of vocational training to military technicians, integrating 65 trades, as well as professional development to facilitate career progression (Rounding et al., 2018). In this regard, one of the training establishments called the Osside Institute,[1] offers four leadership programs to those selected to assume greater professional responsibilities. This training emphasizes communication, decision-making, problem-solving, ethics, and other leadership skills essential for institutional leaders. Overall, the CAF administers more than 1500 courses, every year, for over 40,000 military personnel engaged in training activities. To do so, more than 4000 military and civilian instructors are involved in instructional roles (Rankin et al., 2019). Later in their careers, military technicians may engage in collective and joint training to prepare and enable teams and units to function cohesively, thereby enhancing their operability and effectiveness in combined operations with other government departments and allied military forces.

The *Employment Experience* pillar is defined as "the application and continued development of the knowledge, skills, attitudes, and other characteristics (KSAO)

[1] The actual name of this training establishment is the *Chief Warrant Officer Robert Osside Profession of Arms Institute,* and its mission is "to provide military technicians with the competencies and knowledge necessary to embody leadership that is centered on fostering the highest standards of the Profession of Arms" (https://www.cmrsj-rmcsj.forces.gc.ca/io-oi/io-oi-eng.asp).

previously acquired through education, training, and/or self-development, as well as the acquisition and development of new KSAOs through informal learning, in the performance of assigned roles and duties" (Government of Canada, to be published, section 4.2). As such, technicians will gain experience while they are working at various military bases across Canada and when they are deployed on military operations abroad to obtain institutional and operational experiences. During a military career, when technicians remain employed at the same location for an extended period, they will be assigned to various positions. Otherwise, they will be required to relocate geographically to gain experience in a diversity of work environments. In this respect, technicians will enhance competencies in both the organizational (values and strategies) and functional (technical skills and knowledge) realms.

The *Self-Development* pillar is defined as "organizationally directed or self-initiated, professionally or personally oriented learning that refines or further develops an individual's body of knowledge, intellectual and/or professional skill sets, and attitudes that leads to improving the level of a desired competency or competencies" (Government of Canada, to be published, section 4.2). These self-development activities, whether guided or personal, are typically independent from formal professional development programs. However, the CAF offers several online modules that military technicians can complete on the Defence Learning Network (DLN), in addition to the Canada School of Public Service (CSPS)'s learning catalogue. The technicians can also seek developmental opportunities outside of the government structure to expand their knowledge and skills within their trade.

Considering the ever-changing and evolving security environment worldwide, military technicians must invest time and energy into the four pillars described above to develop, maintain, and enhance their professional competencies throughout their careers. They should also consider acquiring new knowledge and developing new skills to facilitate their eventual transition at the end of military service. Nevertheless, the remainder of this chapter will focus on activities related to the *Training* and *Self-Development* pillars since they refer to vocational training and professional development, with a view of better understanding how academic integrity is embedded within these programs.

Military Ethos and Academic Integrity

In this second section, we will discuss the fundamental values related to academic integrity to compare them with the values of the Military Ethos. We will first put emphasis on the value of integrity but will also address other values that may be conceptualized in a similar or disparate fashion, depending on the learning environment. As such, we will pay particular attention to the value of courage that develops in environments where it is challenged, which can take on different meanings when it comes to a military setting.

The Value of Integrity in Military Contexts

In a military context, a set of values is paramount to guide the conduct of the Profession of Arms, whose "fundamental purpose is the ordered application of lawful force in accordance with government direction" (CDA, 2009, p. 4). Military technicians first join the Profession of Arms before they obtain a trade qualification and are expected to always adhere to the Military Ethos (on and off duty) (CDA, 2024). In 2022, the CAF published a revised iteration of its Military Ethos entitled *Trusted to Serve*, which encompassed six values: loyalty, integrity, courage, excellence, inclusion, and accountability. The value of integrity is defined as "a commitment to a principled approach to duty and being accountable for both action and inaction, and their respective consequences" (CDA, 2022, p. 24). The publication goes one step further to describe what would constitute the actions of "a person of integrity" by stating that it requires "transparency of motive in decisions and actions" and "a dedication to fairness and justice" (CDA, 2022, p. 24). In addition to a set of values, the Military Ethos includes ethical principles and professional expectations, guiding military members in their decision-making so that the Federal Government and Canadian citizens can trust those employed in their military organization.

Given that the military is a collective profession, where individuals cannot perform their duties alone, teamwork is essential for operational effectiveness. Although unique contributions are recognized, they are typically shared with the group with a view of combining talents to strengthen the organization. In that sense, military members are not as keen as individuals evolving in typical academic environments to attribute ideas to their original sources. This way of singling out individuals within the group may be perceived as weakening the team synergy, putting the overall mission success at risk (CDA, 2007). This mindset is instilled in military members during BMQ, right after recruitment, and is also enforced in vocational training and professional development programs throughout their careers. Consequently, military learners may find it more difficult to grasp the concepts related to copyrights and be reluctant to apply such practices in their written products to truly adhere to the basic principles of academic integrity. Furthermore, military members will often reside together during vocational training so that they are located near the training establishment to maximize their learning experience. This colocation of learners is a fertile ground for collaborative approaches to learning activities and assessments, be it encouraged or not by instructors.

The Value of Integrity in Academic Contexts

In 1992, the Center for Academic Integrity (CAI) was established in the United States, effectively launching what has been known as the "academic integrity movement" (Davis et al., 2009, p. 28). This organization became critical for stakeholders to explore academic integrity as an area of research. Over the following two decades, significant development was observed in this field of interest from both practitioners' and researchers' perspectives. In 2010, the CAI was renamed the International Center for Academic Integrity (ICAI) to highlight the importance of the matter in educational settings world-wide (Eaton, 2021). To celebrate its 20th anniversary, the ICAI disseminated its third edition of *The Fundamental Values of Academic Integrity* "to help build a culture of academic integrity in the classroom, at the institutional level, and beyond" (ICAI, 2021, p. 3). To do this, educators, students, staff, administrators, and researchers must commit to the following six values: honesty, trust, fairness, respect, responsibility, and courage. These values serve to inform and improve ethical decision-making capacities to transform ideals into actions in academic communities. As such, Macfarlane et al. (2014) specify that academic integrity "encompasses the values, behaviour, and conduct of academics in all aspects of their practice" (p. 339). Therefore, any breaches of academic integrity could also mean a deviation from one or several fundamental values.

Although the ICAI does not specifically mention vocational training in its publication, it is implied, since the organization promotes academic integrity in educational environments and society at large. A wider scope than just primary schools, high schools, colleges, and universities is important because the literature attests to higher odds of cheating in the workplace when cheating took place during the academic journey (Graves, 2008). It is also recognized that developing and maintaining integrity is challenging (Dyer, 2010). This is why it is so important to engage in the promotion of academic integrity in all learning environments by instilling values that will contribute to the prevention of plagiarism and cheating.

A Comparison of Values Between Academic and Military Contexts

As illustrated in Table 11.1 below, it is possible to draw discernable parallels between the CAF Military Ethos values and the ICAI fundamental values of academic integrity. On both sides of the table, the values have been numbered to show the position in which they appear in their respective reference documents. We matched values by pairs, based on their resemblances, and ordered them by priority of importance when applicable to military vocational training and professional development. As previously mentioned, one of the six values in the Military Ethos is *Integrity*, and although we matched it with the value of *Trust*, its description also presented similarities with other academic integrity values. The elements that are common to both sets of values are identified by an asterisk (*) and are discussed in greater detail in the following paragraphs.

In terms of similarities, both sets of values insist on the importance of setting an example, especially for educators (*Fairness, Responsibility*) or leaders (*Integrity, Loyalty, Excellence*). On both fronts, it can be done by modelling acceptable

Table 11.1 Comparison between the military ethos values and academic integrity values

Military Ethos—Trusted to Serve	Canadian Armed Forces (CAF)	International Center for Academic Integrity (ICAI)	Fundamental Values of Academic Integrity
2- Integrity	Decisions in line with values* Adhere to ethical principles Act on what is right Transparency of motive* Never use role for personal gain	Produce genuine work Build on actions, not words Share information/new ideas Promote transparency* Apply standards equitably	2- Trust
3- Courage	Overcome fear/pain* Resist peer pressure Speak out on wrong-doing* Learn from mistakes* Admit to needing help	Withstand fear/difficulty* Hold all to high standards Address wrong-doing* Make choices, learn & grow* Take risks to defend integrity	6- Courage
1- Loyalty	Leaders set example Obey orders/commands Mutual respect (peers)* Conduct reflects institution Challenge misbehaviours	No compromise on values Value others' opinions Reciprocal respect* Foster active engagement Provide honest feedback	4- Respect
6- Accountability	Accountable for (in) action Share insights with others Stand up to scrutiny Question or correct others Accept consequences*	Safeguard values/principles* Distribute power for change Inspire others to engage Seek to understand rules Enforce institutional policies	5- Responsibility
5- Inclusion	Create sense of belonging Develop unique potential Represent diversity Encourage creativity Challenge inequities	Develop just policies Provide clear expectations Produce original work Evaluate impartially Apply rules consistently	3- Fairness
4- Excellence	Pursue strength of character Healthy competition Lifelong learning Mentor and coach others Be curious/ask questions	Be truthful Give credit to sources Keep promises Provide factual evidence Aspire to objectivity	1- Honesty

behaviours, such as citing sources when teaching or challenging misconduct when it occurs. Additionally, the reciprocal nature of relationships is highlighted by both sides (*Respect, Loyalty, Trust*), in a sense that mutual respect is expected not only between educators and students but also between peers. As such, individuals are encouraged to express their opinions, and institutional representatives are expected to apply regulations equitably. Moreover, transparency of motivations (*Integrity*) and actions (*Trust*) is promoted in both settings, going one step further in the Military Ethos with inactions that must also be justified (*Accountability*). Finally, respect for rules and regulations is key (*Fairness, Responsibility, Loyalty, Accountability)* as long as the standards have been clearly communicated and understood by all involved. In the face of consequences (*Respect, Integrity, Courage*), these should be accepted as an opportunity to learn from mistakes and share insights with others.

Regarding differences, the ICAI set of values encourages the production of authentic and creative work (*Fairness, Inclusion, Trust*) while insisting on the importance of referencing sources of information to attribute credit to original authors. Moreover, the evaluation process is specifically addressed (*Trust, Fairness*), requiring educators to provide clear expectations and evaluate impartially. As for the CAF values, they encourage healthy competition (*Excellence*) between military members by challenging each other to reach higher collective standards. In that respect, no personal gain should be obtained, especially not by those in leadership positions (*Integrity*). Finally, the value of *Inclusion* presents interesting perspectives to increase the representation of diversity and challenge inequities inside smaller groups or the entire organization.

The value of *Courage* is worth its own analysis since it was only added to the ICAI set of fundamental values of academic integrity in 2021, but it has always been part of the CAF Military Ethos. In the two sets of values, it is recognized that courage is required to withstand or overcome fear, as well as difficulty (ICAI) or pain (CAF). All individuals are also encouraged to address wrongdoings by speaking out, which may be challenging, but for different reasons in military or academic settings. In a military organization, members tend to follow a code by which protecting each other is paramount, making it more difficult to report a peer (Shorey, 2000) or a superior at the risk of being disloyal. This is most likely why the Military Ethos stipulates to "resist peer pressure," which is easier said than done in a collective culture. In a typical academic environment, students are also hesitant to "snitch" on their peers at the risk of being ostracized by the group (Rangkuti et al., 2022) or on their professors for fear of retribution. One last similarity between the two sets of values is learning from mistakes after making choices. To avoid mistakes, an interesting element of the Military Ethos is to admit needing help, which does require adopting a vulnerable position. All in all, "courage can only develop in environments where it is tested" (ICAI, 2021, p. 10). From an academic integrity perspective, we wondered whether learners in a military vocational context would perceive themselves as courageous if they submitted original (but perhaps flawed) work, taking the risk of earning lower grades.

To embody a set of values, these values need to be developed, practiced, and maintained. Although military members mostly adhere to the Military Ethos in non-academic environments, they still find themselves partaking in vocational training and professional development throughout their careers. Therefore, we thought that it would be interesting to explore how Military Ethos values, especially the one of integrity, are embedded into academic settings within the military organization.

CAF Vocational Training and Academic Integrity

The RMCC in Kingston defines academic integrity as "when you do the right thing, even if no one is watching" (RMCC regulations, article 23). This military college represents the perceived authority for the development of academic integrity policies for the entire CAF. When it comes to vocational training, there is a general lack of academic integrity policies being enforced by the establishments where military technicians obtain trade qualifications. This may be justified by the nature of the learning activities that are highly practical, such as using a weapon or fixing a vehicle. Another reason would be that academic misconduct definitions and processes are based on Defense Administrative Orders and Directives (DAOD), partially tailored to fit a post-secondary institutional context. These directives are also vetted by the Judge Advocate General (JAG) since the CAF operates its own internal justice system. Therefore, investigation processes on academic integrity breaches may not be dealt with only by granting academic sanctions but also by imposing formal "administrative or disciplinary" measures that may impact a military member's career progression. For example, when a military technician is caught cheating or plagiarizing during vocational training, a Progress Review Board (PRB) composed of leaders, instructors, and teaching standards experts will decide on the severity of "administrative" sanctions (i.e., cease training, return to home unit, remedial training, etc.). In more extreme cases, a "disciplinary" sanction will be considered, such as a reprimand, a fine, or minor punishments, following a summary trial.

Unlike the United States military training institutions, the CAF does not have honour codes that would require or encourage peer reporting (McCabe et al., 2001). Even where these codes are enforced, the organizational culture plays a larger part in determining "attitudes toward academic integrity rather than the presence or absence of a code" (Eaton & Fishman, 2024, p. 178). In the training context we are examining for this chapter, there are limited expectations toward learners and instructors to respect academic integrity principles besides their usual adherence to the Military Ethos values, ethical principles, and professional expectations. However, when military technicians are selected to attend leadership programs delivered by the Osside Institute, this training institution does have an academic integrity policy, in part because the establishment is a subset of the CMR in Saint-Jean. Although the policy is included in the students' learning manual, it is seldom discussed between the instructors and learners, especially during distance learning programs. Moreover, when it comes to plagiarism, it is based on a strict application of referencing standards, such as the ones from the American Psychology Association (APA), that are not explicitly taught to students. Since the majority of military technicians did not complete postsecondary education and do not use such referencing styles during their regular work, they are bound to make mistakes in their assignments, which would be considered as unintentional plagiarism (Jamieson & Howard, 2019). The Osside Institute has recently revamped some of its leadership programs to steer away from asking learners to write academic papers (i.e., essays). It is now focusing on teaching military technicians advanced military writing skills that include referencing sources of information, but in a more simplistic format. For example, a telephone conversation could be referenced as: "Telcon between Major Smith and Captain Brown dated 25 July 2023." In this case, military technicians can put into practice transferable skills by referencing sources of information commonly used in their workplace, such as government documentation, internal reports, and even points of contact within their professional network. Additionally, based on expert recommendations, learners are now encouraged by instructors to start using GenAI to complete assignments during professional development programs.

Artificial Intelligence and Military Training

The use of AI by military forces across the world is not new. This type of technology has long been integrated into sophisticated target detectors (radars and sonars) and weapon systems (drones). It is also useful to predict (analytics) when equipment will require maintenance or to recognize (sensors) failures in various systems. Regarding training, AI has been used in simulations for virtual wargaming and virtual reality (VR), as well as "digital twins" to represent physical objects and systems. Given that military operations present risks to human safety, rights, and privacy, military technicians must learn how to ethically use AI to identify and address harmful situations. Although the principles of AI ethics go beyond academic integrity, they are key to shaping conduct and behaviours in learning environments. Therefore, in this last section, we will first look at the more recent developments with respect to AI government guidelines. Then, we will discuss how these guidelines currently influence or will impact the use of AI in military vocational training and professional development.

Artificial Intelligence in Government Policies

In early 2024, the DND and CAF published an *Artificial Intelligence Strategy* to "provide a vision and direction for the development and integration of AI and automated decision systems" by 2030 (Government of Canada, 2024a, p. 1). In this document, AI is defined as "the capability of a computer to do things that are normally associated with human cognition, such as reasoning, learning, and self-improvement" (p. 4). A quick search in the strategy document reveals that the theme of "education" is not at the forefront of the policy. However, military members are encouraged to review and adapt training programs "to incorporate AI-related skills and to address issues such as the ethics and military implications of data and advanced technology" (p. 21). To do this, the CAF intends to use curricula that are both developed internally and procured externally, as well as establish academic partnerships. In a military context, there is a greater focus on "general" AI technology than "generative" AI tools. However, the Artificial Intelligence Strategy does include a section on GenAI that identifies it as "a subset of Machine Learning which can produce a wide variety of novel content, such as images, videos, audio, text, code, and 3D models, in response to user prompts" (Government of Canada, 2024a, p. 4). It is interesting to note that "text" ranks fourth in the type of content that can be generated by AI, when it is the content of greater concern for those promoting academic integrity in most educational settings.

In February 2024, the Government of Canada also published a *Guide on the Use of GenAI* that includes policy considerations and best practices for all federal departments, while drawing attention to biases in the training data and information security requirements (Government of Canada, 2024b). To ensure the ethical use of GenAI, the guide highlights the principles of accountability for the impact of content generated by AI, as well as transparency by identifying content produced by interacting with a GenAI tool. All federal employees are also encouraged to educate themselves about GenAI by learning about its strengths and limitations to mitigate risks to the public. Even before these guidelines were published, the RMCC updated its academic regulations to deal with the impact of GenAI for the benefit of professors and students. It was decided that there would not be a blanket policy prohibiting the use of GenAI tools but that educators would maintain academic freedom by identifying their level of integration when teaching.

As for GenAI detection tools, such as *GPTZero* and *Turnitin*'s GenAI detection plugin, professors were strongly advised against their use, in part due to their lack of reliability. In general, the use of text-matching software, such as *Turnitin*, can be controversial because of the security concerns associated with the retention of student assignments in global databases. Not only do the content of these papers become accessible by external parties, but also the metadata of these documents, which could constitute a violation of copyrights and privacy (Foster, 2002). In the case of military colleges, this represents greater security concerns if the assignments are based on references that include sensitive information not accessible in open sources. At this time, proving the use of GenAI remains challenging, which may result in students being wrongly accused of plagiarism (Carter, 2023) or entice students to circumvent usage restrictions of various tools. To that extent, it is recognized that the pressure to succeed at RMCC has motivated students to develop deference to authority, which is reflected by an altered motto: Truth, Duty, Valour "and don't get caught" (Arbour, 2022, p. 226). These students will occupy leadership positions in the CAF upon graduation and supervise military technicians during their careers. It then raises the question of whether they would take future academic breaches of tradespersons seriously if they had adhered to lower standards of integrity during their own academic journey.

Artificial Intelligence in Vocational Training and Professional Development

In spite of the recent revision of RMCC academic integrity policy, there is no indication that establishments delivering vocational training or professional development programs have followed in the same footsteps. Even though GenAI has been a major topic of discussion across formal educational institutions across the globe, the CAF vocational training establishments do not seem to be sharing the same concerns. As stated in the DND/CAF Artificial Intelligence Strategy, it is paramount that military technicians be exposed to AI in general (including GenAI) during vocational training so that they develop modernized skills that will be required in future security environments. This orientation is in line with Hennessy (2023), stating that AI will not replace all workers but that those who fail to use AI tools effectively may be replaced by those who can. This highlights a crucial role for educators to not dismiss GenAI as a mere nuisance but to educate students in using GenAI and other AI tools expertly, critically, and ethically. This creates a window of opportunity for vocational training establishments to reflect on the ethical repercussions of failing to use AI in a responsible manner while military technicians work toward a specialization in operating sophisticated equipment and systems. Ultimately, breaches in academic integrity during vocational training may result in military technicians not having acquired the necessary knowledge or developing the appropriate competencies. The consequences of this can be as severe as putting the lives of human beings in jeopardy.

Regardless of the means used by learners to plagiarize or cheat, academic integrity in military training is currently regulated by institutional directives, with an emphasis on detecting academic misconduct and sanctioning these failures. However, there is an opportunity to further develop policies that would be tailored to the technical and practical nature of vocational training in a military context, which could enhance the effectiveness of training programs and uphold higher standards of professionalism and ethical behaviour. However, given the growing academic integrity challenges faced by learners in the digital and AI eras, the organization would benefit from developing and implementing strategies for preventing plagiarism, promoting academic integrity, and incorporating restorative justice alternatives to address academic misconduct constructively (Sopcak & Hood, 2022). Such a comprehensive approach could not only dissuade academic dishonesty but also foster an overall ethical culture within military vocational training and professional development programs.

Conclusion

There is still much work that remains to be accomplished regarding academic integrity in the CAF military vocational training and professional development programs. Considering that the Military Ethos values are closely connected to those of academic integrity, there seems to be an assumption that military trainees are less likely to engage in academic integrity breaches. Moreover, the fact that vocational training usually focuses on practical tasks makes it more difficult to understand how conventional plagiarism or cheating definitions can apply to the learning environment. Lastly, the collective nature of military service influences how academic integrity is conceived, attributing a lesser degree of importance to individual or original ideas. Therefore, not all training establishments have developed separate academic integrity policies specific to vocational training, especially in this new GenAI era. Instead, they rely on administrative and disciplinary regulations that could apply to non-compliance with military values in any situation, be it when military technicians are working or training.

Looking at the future, it would be worthwhile for the CAF to consider how Military Ethos values specifically apply to vocational training for the development of academic integrity policies. It would also be conceivable that the first instances of unintentional academic misconduct were addressed outside of the regular administrative and disciplinary systems. This would allow military technicians to learn from their mistakes in training environments without fearing repercussions on their military career. The use of restorative justice mechanisms could be contemplated, such as remedial training, peer briefings, and apologies to instructors, to explore how harms can be repaired (Karp, 2019). These constructive strategies would empower military technicians to rebuild trust through active accountability, which is aligned with the Military Ethos.

Finally, further research could be done on academic integrity and military vocational training by exploring the types of AI used by the different trades. This information would provide better indications on potential academic breaches by military technicians using sophisticated technologies. By developing a better understanding of learners and learning situations, tailored academic integrity policies could be implemented, with a view to prevent plagiarism and cheating in vocational training and professional development programs.

References

Arbour, L. (2022). R*eport of the independent external comprehensive review of the Department of National Defence and the Canadian Armed Forces*. http://central.bac-lac.gc.ca/.redirect?app=damspub&id=3a6343e9-fc51-4519-8f70-b87d269eb565.

Bériault, K.A. (2021). *Retention in the Canadian Armed Forces: A complex problem that requires active and continued management*. [Major Research Project, Canadian Forces College]. https://www.cfc.forces.gc.ca/259/290/23/286/Beriault.pdf.

Canadian Defence Academy. (2007). Leadership in the Canadian Forces: Leading People. https://publications.gc.ca/collections/collection_2013/dn-nd/D2-313-3-2007-eng.pdf.

Canadian Defence Academy. (2009). Duty with honour: The profession of arms in Canada. https://publications.gc.ca/collections/collection_2022/mdn-dnd/D2-150-2009-eng.pdf.

Canadian Defence Academy. (2022). Canadian armed forces ethos: Trusted to serve. https://www.canada.ca/en/department-national-defence/corporate/reports-publications/canadian-armed-forces-ethos-trusted-to-serve.html.

Canadian Defence Academy. (2024). Fighting Spirit: The profession of arms in Canada. https://www.canada.ca/en/department-national-defence/corporate/reports-publications/fighting-spirit.html.

Carter, T. (2023). Some universities are ditching AI detection software amid fears students could be falsely accused of cheating by using ChatGPT. Business Insider. https://www.businessinsider.com/universitiesditch-ai-detectors-over-fears-students-falsely-accused-cheating-2023-9

Davis, S. F., Drinan, P. F., & Bertram Gallant, R. (2009). *Cheating in school: What we know and what we can do*. Wiley-Blackwell.

Dyer, K. (2010). Challenges of maintaining academic integrity in an age of collaboration, sharing and social networking. In Proceedings of TCC 2010 (pp. 168–195). TCCHawaii. https://www.learntechlib.org/p/43770/.

Eaton, S. (2021). *Plagiarism in higher education: Tackling tough topics in academic integrity*. Libraries Unlimited.

Eaton, S. E., & Fishman, T. (2024). Academic integrity in North America: A comparison of Canada and the USA. In S. E. Eaton (Ed.), *Second handbook of academic integrity* (pp. 172–184). Springer.

Foster, A. L. (2002, May 17). Plagiarism-Detection Tool Creates Legal Quandary. The Chronicle of Higher Education. https://www.chronicle.com/article/plagiarism-detection-tool-creates-legal-quandary/

Government of Canada. (to be published). Defence Administrative Orders and Directives 5031-8: Canadian Armed Forces Professional Development. Department of National Defence.

Government of Canada. (2024a). The Department of National Defence and Canadian Armed Forces Artificial Intelligence Strategy. https://www.canada.ca/content/dam/dnd-mdn/documents/reports/ai-ia/dndcaf-ai-strategy.pdf.

Government of Canada. (2024b). Guide on the use of generative artificial intelligence. https://www.canada.ca/en/government/system/digital-government/digital-government-innovations/responsible-use-ai/guide-use-generative-ai.html.

Graves, S. M. (2008). Student cheating habits: A predictor of workplace deviance. *Journal of Diversity Management, 3*(1), 15–22. https://doi.org/10.19030/jdm.v3i1.4977

Hennessy, A. (2023, September). Learn AI now or risk losing your job, experts warn. CBC News. https://www.cbc.ca/news/canada/artificial-intelligence-jobs-careers-training-panle-the-national-1.6978515.

International Center for Academic Integrity (ICAI). (2021). *The fundamental values of academic integrity.* (3rd ed.). https://academicintegrity.org/images/pdfs/20019_ICAI-Fundamental-Values_R12.pdf.

Jamieson, S., & Howard, R. M. (2019). Rethinking the relationship between plagiarism and academic integrity. *International Journal of Technologies in Higher Education, 16*(2), 69–85. https://id.erudit.org/iderudit/1067061ar

Karp, D. R. (2019). *The little book of restorative justice for colleges and universities: Repairing harm and rebuilding trust in response to student misconduct* (2nd ed.). Good Books.

Macfarlane, B., Zhang, J., & Pun, A. (2014). Academic integrity: A review of the literature. *Studies in Higher Education, 39*(2), 339–358. https://doi.org/10.1080/03075079.2012.709495

McCabe, D. L., Trevino, L. K., & Butterfield, K. D. (2001). Cheating in academic institutions: A decade of research. *Ethics & Behavior, 11*(3), 219–232. https://doi.org/10.1207/S15327019EB1103_2

Rangkuti, A. A., Royanto, L. R. M., & Santoso, G. A. (2022). Ethical awareness and peer reporting intention of exam cheating and plagiarism: Mediation role of ethical judgment. *Issues in Educational Research, 32*(3), 1111–1130. https://www.iier.org.au/iier32/rangkuti.pdf

Rankin, K.J., Rounding, K., & Ford C. (2019). Military instructors: Understanding motivation and job satisfaction across two nations [Report #DRDC-RDDC-2019-R201]. Director General Military Personnel Research and Analysis.

Rounding, K., Ruscito, F., & Rankin, K. J. (2018). *Canadian Armed Forces instructors satisfaction and dissatisfaction*. Defence Research and Development Canada.

Shorey, G. (2000). Bystander non-intervention and the Somalia incident. *Canadian Military Journal, 1*(4), 19–28.

Sopcak, P., & Hood, K. (2022). Building a culture of restorative practice and restorative responses to academic misconduct. In S. E. Eaton & J. Christensen Hughes (Eds.), *Academic integrity in Canada: An enduring and essential challenge*. Springer.

Catherine E. Déri, Ph.D., is a Associate Professor in the Department of Education Sciences at Université du Québec en Outaouais, and an Adjunct Professor in the Faculty of Education at University of Ottawa. The topic of academic integrity piqued her interest during her doctoral studies on the professionalization of Ph.D. students to the scholar profession. She currently conducts research on plagiarism prevention in academia through an international partnership involving 70 researchers in Canada, the United States and Europe. Her transition into the field of higher education comes after a 25-year career as a Logistics Officer in the Canadian Armed Forces. Her honours include the Pierre Laberge Thesis Prize from University of Ottawa, the 10x10 Alumni with Impact Award from University of Guelph, and the Robert J. Farrelly Award from the Royal Military College of Canada.

Caroline Kirouac is a Ph.D. Candidate in the Department of Education Sciences at Université du Québec à Rimouski (Campus Levis). Since 2014, she has been employed as a Training Development Officer with the Canadian Armed Forces, providing guidance on the development of training programs within the military education system. She is currently conducting research on the use of technopedagogy (simulators) to enhance the self-efficacy of military instructors and Training Development Officers during vocational training. While completing a doctoral internship, she joined the international Partnership on University Plagiarism Prevention at Université du Québec en Outaouais, where she studied the locus of control of college teachers from an academic integrity perspective.

Lucie Moussu, PhD, holds the positions of Professor in Writing Studies, Director of the Royal Military College Writing Centre, and Chair of the university's generative AI Committee. From 2009 to 2021, she was Associate Professor in Writing Studies and Director of the University of Alberta's Centre for Writers. Prior to this, she worked at Toronto Metropolitan University for 3 years. Dr. Moussu is a member of the Journal of Response to Writing Editorial Board, and she also regularly serves on SSHRC Grant Adjudication Committees. Her research interests include Second Language Writing, writing centre administration, and the role of generative AI in shaping pedagogy and academic integrity policies.

Part III
Educators Support of Academic Integrity: Vocational Training Vignettes

Chapter 12
Student Engagement in Geomatics Engineering Technology Assessments

Carina Butterworth

Abstract As a detailed and technical profession, integrity is integral to geomatics engineering technology. This field is focused on the use of technology for measurement and communication of those measurements. Satellites, airborne, and terrestrial platforms are used to measure the earth's surface or within the earth's surface to develop accurate models and deliverables of high accuracy to a client. A geomatics engineering technology instructor in a Canadian polytechnic institution shares how assessments that are designed to engage student passion will empower learning about a topic over aiming for a grade.

Several successful techniques have emerged from practice to help increase engagement in the material. One technique is providing students with choices in project scope, grade weighting, and submission format. A second technique focuses on interpretation and analysis rather than the process. While the process can be taught to get students to a result, the soft skill of communicating what they did and what the result shows provides a clearer assessment of learning. A third option is zero-risk submission, where the student can submit an infinite number of times until they are happy with their result or a deadline is met. Using these techniques, the author shares observations that students exhibit higher levels of engagement and improved academic integrity in their learning.

Keywords Academic integrity · Learner motivation · Learner engagement · Assessment design · Student-centered learning · Instructional approach · Geomatics engineering technology · Canada

C. Butterworth (✉)
Southern Alberta Institute of Technology (SAIT), Calgary, AB, Canada
e-mail: carina.butterworth@sait.ca

L. Vogt, B. M. Stoesz (eds.), *Academic Integrity in Vocational and Polytechnic Education*, Ethics and Integrity in Educational Contexts 8,
https://doi.org/10.1007/978-3-031-90091-4_12

Introduction

Student-centred learning has been a focus of educational philosophy for decades (Dewey, 2004), and students are encouraged to be more autonomous with their work and take more control of their own learning (Jones, 2007). Instructors can apply these philosophies to tap into students' intrinsic motivation and increase student engagement in the course material (Barnes, 2013; Martin, 2008). The time and effort a student engages with learning are positively correlated with the mastery of skill and knowledge of that material (Drenth et al., 1998). Students will exceed expectations and want to pursue new knowledge when their passion and excitement are triggered (Hart et al., 2011).

Academic integrity is the responsibility of both the instructor and the student (Chirikov et al., 2020; Eaton, 2021). Course and assessment design are integral in implementing academic integrity principles for engaging students (Hart et al., 2011). Wiggins (1989) found that a limited style of assessment is demotivating and disengaging to students; therefore, Tai et al. (2023) and Hockings (2010) recommended more inclusive assessment designs that allow for flexibility in assessment style and student choice. Several techniques are recommended in current research to assist with assessment design, and this chapter expands on three approaches: student choice in assessments, authentic assessment implementation, and no-risk formative feedback.

The first available technique in assessment design is to provide choices to students on how they will be evaluated. The choices could include which assessments they will complete (within certain restrictions), how they will be graded, and how they would like to complete the assessment for submission. The 'how' in completing this submission is entirely up to the student, including, but not limited to, presentations, posters, videos, game design, or comic strips. This technique applies the mitigation approach described in Hodgkinson et al. (2016).

Authentic assessment is also a technique that has been shown to be effective in engaging students and promoting autonomous learning (Bernhardsson, 2023; Butterworth, 2024; Paull et al., 2019). Wiggins (1989, p.703) defined authentic assessment as "a true test" of achievement because it replicates real-world problems and requires a developed skillset and knowledge that a professional would use. In authentic assessments, students are evaluated on *doing* rather than *knowing*. This approach provides students the opportunity to communicate what their actions and results mean, which makes tools such as generative artificial intelligence (genAI) difficult to use when the assessment is based on unique scenarios. The emphasis on communicating results and analysis will be discussed as a second technique, but it is an extension to providing choices in an assessment.

The third technique discussed is the power of formative, no- or low-risk assessment in academic integrity (Eaton et al., 2017). When students feel safe to submit work for editing or feedback, they are less likely to plagiarize. Using these three instructional and assessment strategies, the instructor actively shapes the course design and assessment techniques, aligning them with the students' abilities. This approach, along with the effectiveness of the assessments in evaluating students' learning, influences student motivation to uphold academic integrity.

Background

Geomatics Engineering Technology

The Geomatics Engineering Technology program at the Southern Alberta Institute of Technology (SAIT) in Calgary, Alberta, Canada, is the program of interest in this chapter. It is a two-year diploma program where students learn specializations in measurement and analysis. Leading-edge technology in laser scanning, unmanned aerial vehicles (UAVs), global navigation satellite systems (GNSS), land surveying techniques, and geographic information systems (GIS) are examples of courses taught over the four semesters, with five courses per semester. The students are in class for between 25 and 32 h per week, depending on the semester, with over 300 h of field work laboratory time throughout the program. There are 64 students accepted each year, with a maximum of 128 students between the 2 years.

The course material is developed based on industry needs and modified as the industry requires new skills or knowledge, but the assessments are designed by the instructors. A program advisory committee (PAC) consists of 12 industry members, and all the Geomatics Engineering Technology faculty members assist in developing the course curriculum and assessments. The committee meets once a year and spends time in groups working with the instructors of specific course materials. The instructor is informed of industry changes and needs, and the industry members provide feedback on the validity and practicality of projects and assessments.

Teaching Philosophy

This chapter is written with a radical constructivism teaching philosophy (von Glasersfeld, 1995). The Geomatics Engineering Technology program design follows a social constructivist philosophy, and the instructor of these courses implements elements of the program philosophy with the radical constructivism approach. The radical constructivism philosophy puts the instructor in a position as a guide rather than a keeper of knowledge. It comes with a perspective that students know the material from their previous experiences and only need assistance constructing information into a form they can understand and communicate. The first core principle of radical constructivism is that reality and learning are adaptive and integrated. There is no separation between one's reality and the construction of knowledge. Engaging with the material in ways that mean something to the student inspires them to learn more and take more risks (Glowatz & O'Brien, 2017). This is the second principle of radical constructivism, where knowledge must be built up through active work (Walshe, 2020). These approaches also complement social constructivism in the program, where the teaching approach is focused on creating collaborative and socially focused learning designs (Vygotsky, 1978).

The second principle also relies on the belief that education is a community where relationships are essential to learning (Fulford, 2021). This relationship can be defined as a pedagogical partnership (Cook-Sather et al., 2023). In a safe community of people, students are able to discuss their previous experiences and learnings. The students will also express their experiential reality (or otherwise described as their experience of everything) with each other, creating an environment where students can confirm their learnings and collaborate on building knowledge (von Glasersfeld, 1991). The independent construction of the material and the social component of learning are complementary and instrumental (Cobb, 1994) in humanizing their learning and the student-instructor relationship (Cook-Sather et al., 2023). The design of assessments in the radical constructivism philosophy uses guidance, relationships, and engaging assessments from the core values of radical constructivism. This approach is often uncomfortable for the students at first, but the commitment to academic integrity and improvement in performance after implementing this philosophy is noticeable.

Implementation of Academic Integrity: Supporting Techniques

Choice-Based Assessment

One of the techniques used in teaching Geomatics Engineering Technology courses is the option for students to make their own choices in how the course will progress. Courses of interest were Mapping Fundamentals in the first semester, Remote Sensing in the third semester, and 3D Modelling and Capstone projects in the final semester.

First Semester: Mapping Fundamentals Limited choices are offered in the first semester due to the students being overwhelmed with the transition from high school or industry to a post-secondary environment. The Mapping Fundamentals course covers a diverse range of subjects, from team and organizational skills to calculating magnetic declinations. Originally, the students were given the option to choose the grading scale for all their assessments in this course and when the assessments would be due. However, because the students have limited experience in post-secondary institutions, these choices seemed to create confusion, resulting in students performing poorly or deciding not to submit their self-assigned low-risk assessments.

The choices were adjusted so students could decide how much the pieces of a single assessment were worth. This assessment is called the "Term Activities Assessment" and is worth 25% of the final grade. It is set up with hands-on activities for the students to complete during class, with a learning reflection they complete at the end of each topic. The reflection can be created in any way the student likes, provided they capture what material stood out to them and how important they found it. Submissions for the reflection have come in the form of "Dear Diary," videos, audio clips, drawings, and the traditional answers to questions. The in-class activities are often performed in groups, so the reflection provides the instructor insight into what the student remembers from the group activity. The students choose what each topic should be worth. They are given the assessment and access to all the course material in advance to assist with their choice (Eaton & Kumar, 2023), but with the expectation that the decision is submitted to the instructor by the end of the first week of classes.

Providing options and flexibility to the students in their first semester teaches them how to start organizing their priorities, which contributes to a culture of academic integrity (McCabe et al., 2001). The reflection is difficult to plagiarize because of the inherent nature of reflecting on one's learning progress. The requirements are to include what the student learned and how they feel it is important to them and their career. A few examples of prompts for the reflection are:

- What is your goal for learning in this module? Remember to be specific, measurable, attainable, realistic, and timely.
- List four objectives that you learned in this module.
- Explain which in-class or out-of-class activity helped with each of the four things you learned.
- Why do you believe you were required to learn these four objectives?
- What did you find surprising in this module?
- What do you need to review again? Do you have any questions that still need to be answered? What are they?

This part is challenging to artificially create using genAI (Eaton, 2023), through contract cheating (Eaton, 2022), and by copying other students because it is submitted in a video or audio format. It is also completed during class time, during which the instructor engages in conversation with them throughout the process of creating the reflection. It also brings a sense of the learning process to the students in their first semester of the program (Hodgkinson et al., 2016). The in-class activity submissions are often the same between students because of how they are developed in terms of team assessments, and it is explained at the time the students begin the activity that it is expected they will be the same. This expectation allows students to engage in the learning activity more because they are not worried about the end product as much (Hodgkinson et al., 2016).

Third Semester: Remote Sensing In Butterworth (2024), authentic assessment in this remote sensing course was discussed at length, but academic integrity was missing in the discussion. The course material is designed as a workflow through the physics fundamentals of earth observation imaging to the product of detecting change on the earth's surface. Satellite imagery is used and analyzed throughout the course. The students learn the concepts of the algorithms and then apply the algorithms to a project that serves as the foundation of the course. Students choose their own topic for the project based on a location or aspect of the earth they would like to investigate more closely. The instructor suggests large area changes such as climate change effects, natural disasters, or large human projects that affect the surface of the earth. Examples of projects submitted have been volcanic eruptions close to Alaska in the United States of America, sea water level changes in Indonesia, green roof construction in China, and crop health by Stettler, Alberta, Canada. Every year, new projects across the world are chosen for different reasons that are personal to the student. The students take a competitive, scientific approach where they want to prove what they are seeing through imagery. The foundation of remote sensing analysis is that there is information in the numbers, not just in the visualization. The numbers can be processed to create new images that were not seen in the initial image. Another component of remote sensing is that proof of what the images show still needs validation. Images cannot be taken as truth without a way to validate the data.

Students select the images they believe best prove the changes to the earth that are the focus of the project. The instructor suggests Landsat data to help keep data collection simple and ensure students have enough data to apply different algorithms (U.S. Geological Survey, 2023). When it comes to processing the data, the students must search for the ideal methods for their projects and incorporate the information they have learned in class. Before ChatGPT was a common tool in education, one student said, "I can't just Google the answer because it is not there," demonstrating that students must make their own decisions and find innovative solutions. Also, the project is submitted as a report with all the raw and processed data, so the data keeps the student engaged in creating results independently. Students have shown a higher level of engagement in the course, and there has been a significant decrease in academic dishonesty issues when given these choices, as observed by the instructor. The primary focus of authentic assessment is to evaluate students on what they can do rather than what they know, which keeps academic integrity at the core of assessment design (Janesick, 2006).

Fourth Semester: 3D Modelling and Capstone There are two courses in the fourth semester where students have many high-stakes choices provided to them. The first course is 3D Modelling, where the instructor provides students with a "choose your own adventure," according to one student. Butterworth (2024) and Hummel and Butterworth (2024) describe the authentic assessment components in detail, but the academic integrity lens was not discussed. The 3D Modelling course has a laser scanning focus, where lasers are used to measure the surface of the earth or a human structure, and modelling theory is applied to create digital models. The students have two projects and three exams, and the students can choose the percentage of the final grade each component is worth, how many graded components to submit, and when those parts are to be handed in. There are end dates to the final project, but the pieces of the project due dates are flexible.

The projects have a broad scope for the learning process, but the specific topics are not necessarily defined by the instructor. The first project uses airborne laser scanning (lidar) data to measure and extract features from the provided dataset. One of the measurements is to find average tree heights, but to get there, the student needs to create the intermediate models. A basic workflow is provided, but the student needs to decide what specific parameters to set in the workflow. These parameters are often found through trial and error, and progress is communicated through a cumulative report, in which each step is documented in detail. Each student has a slightly different approach depending on their summer job experience and how they interpret the results. There is a correct answer, but several different ways of getting to it.

The second project in the 3D Modelling course is a fieldwork exercise where the student will measure an apprenticeship lab room with a laser scanner mounted onto a tripod. The student chooses which lab room to measure and must initiate communication with the lab instructor to arrange a time to complete the measurement. Different labs have different protective equipment requirements, so the students should choose which lab they feel represents the industry they would like to work in. The student must also find out what product the lab instructor would like as a deliverable. If that instructor does not understand what types of products the laser scanning project can provide, the student is required to choose a suitable product. The student documents their work in a format they feel is best and provides a final deliverable to the course instructor. The final deliverable can be anything they like, except for a report. Examples of submissions have been webpages, comic strips, newscast videos, board games, children's books, and posters. Figure 12.1 shows the cover of a children's book created for this assignment. By encouraging alternative deliverable formats, students have more control over their project, and they have a chance to have fun at the end of their final semester. This has been shown to engage the students more in the course and showcase their skills in ways they may not have had a chance to throughout the 2 years, improving confidence in submitting their own work.

The second fourth-semester course with choices is the capstone project. Students choose or create a project of interest to them and are entirely responsible for collecting their own data, contacting stakeholders, processing the data, and providing an industry-ready result. The capstone project has less flexibility in timelines and grading components than 3D Modelling, but the projects are entirely within the student's control and decision-making. Instructors act as advisors to help answer questions and provide technical feedback, but they are not teaching the students any material. Because the final deliverables are a comprehensive report and presentation, the students do not have flexibility in their submission, but they have control over the format of the results and how to provide it to any client they have chosen to work with. Academic integrity and the capstone projects are compatible in terms of the ethics of each project. The capstone projects are based on industry projects, and the results

Fig. 12.1 Example of a student-produced children's book. *Note:* This image has been used with permission by the authors. *From* Montiel, C. K., & Basil, B. S. (2024). *The amazing stories of point and cloud*. MAPS362: 3D Modelling. SAIT, Calgary, Canada

are provided to industry. If the student provides a product that shows non-ethical academic data, the impact will be the student's ability to work in the industry (Khaled & Lois, 2019). Generally, academic integrity issues in this project come from the report writing itself, but mitigation techniques will be discussed in the third technique of this chapter.

Interpretation and Analysis Communication

In all the courses, interpretation and analysis are emphasized in the assessment deliverables. Each time a result is created, the student is required to explain what the result shows and how it relates to the scope of the project. The student also needs to justify the steps taken to achieve the result. These are skills that the PAC has requested us to teach in the program, and this industry need is communicated to the students. Academic dishonesty has been countered by emphasizing on interpretation and analysis. With different projects and results in assessments, students cannot rely on each other for an answer, and genAI is not yet developed well enough to analyze the infinite number of scenarios in geomatics. An artificial intelligence program cannot fully and appropriately describe the results of a project, and students have different results, so they cannot use artificial intelligence or copy each other's responses. They also do not require many references to interpret and analyze their results, so citation mistakes are uncommon.

The other reason for doing this is related to the Motivation and Engagement Scale (MES), as Martin (2010) described. Part of the MES wheel includes the adaptive cognition of mastery orientation, and the student describing their results shows they are able to understand and communicate what they have learned in the project. If the student has a well thought out description of the results, this suggests that they value their learning and are engaged in the material, which shows enhanced academic integrity. The students can also use their projects as a portfolio to show their potential employers, so individual engagement is increased as the project outcome is linked to future employment opportunities.

Formative Feedback

Formative assessment is likely the most beneficial academic integrity tool in the Geomatics Engineering Technology program and is a recommended strategy (Black & William, 2009; Pecorari, 2015). The definition of formative assessment, as provided in the Council of Chief State School Officers (2018), is "… a planned, ongoing process used by all students and teachers during learning and teaching to elicit and use evidence of student learning to improve student understanding of intended disciplinary learning outcomes and support students to become self-directed learners" (p. 2). Geomatics Engineering Technology students include high school students who struggled to achieve the required grades to enter a university or mature students making a career change from a trade or unskilled labour position. Students may also speak English as an additional language. Using a radical constructivist philosophy, the instructor works as a guide to the learning process and uses assessments as building blocks to support students in growing their confidence as learners.

Removing high-stakes assessments in favour of formative assessments has improved academic integrity within the courses (Higgs, 2014). The formative assessments are considered no-risk assessments. The student is allowed to submit the assignment as many times as they like for grading and can redo the submission as often as required until they either earn the desired grade or the deadline has been met. In support of the pedagogical partnership, this method supports equitable teaching, where students feel the instructor understands that students have different challenges and provides validation to the students' humanity (Cook-Sather et al., 2023).

In the first semester course, the students begin with small assignments where they answer randomly selected questions from a larger question bank. They can submit these as many times as they like until they get the correct answer. In the third and fourth semesters, the students can send their projects to the instructor for feedback or "red lining" of the project without any consequences (Butterworth, 2024). Consistent with Eaton et al. (2017), formative feedback has improved students' writing and citation skills, reducing the number of academic dishonesty reports and changing the focus away from punitive actions to a learning process. For the instructor, the regular no-risk submissions by the students provide insight into what they are learning and how their skills are developing before assigning grades. This process is difficult to plagiarize as it is very individual and unique to each person. The instructor requires scheduled time to review assessments, which changes the workflow. The instructor of these courses schedules 30–60 min three times a week to review assessments. The students are asked to include a comment in their submission that points the instructor to specific sections or types of material that require focused feedback. If the instructor has time, additional feedback is provided to the student above what they ask. As a result, the final submission is much faster to grade because the instructor has already identified areas for improvement before grading. Additionally, the amount of time dedicated to following up on concerns of academic misconduct post-submission is greatly reduced. Although the upfront work is increased, the instructor experiences time-saving benefits at the end.

Conclusions

Authentic assessment, radical constructivist assessment design, and formative feedback have been instrumental in the development of skilled, ethical, industry-ready students in the Geomatics Engineering Technology program. Three techniques were discussed to promote academic integrity in the program and have been shown to be successful. The first technique allows students to make choices in their assessments and grading to help them feel in control of their work and engage them at a higher level. The second inspires students to focus their efforts on interpreting and analyzing their results, which helps encourage a passion for troubleshooting and communicating their knowledge. Finally, the third technique uses no-risk formative assessment to help students build confidence in their skills and encourage them to take risks in learning more.

As academic integrity and technology change, additional techniques must be investigated, trialed, and reflected on to keep students engaged and passionate about learning. An assessment for the instructor, just like for the student, is a living document that can be changed, modified, and grown until the deadline is met. An assessment is also where design and philosophy merge to bring ethical learning opportunities to all the students in their courses. The assessment must be designed as an extension of a teaching philosophy where students engage willingly in the material, and their passion for learning about the topic is enhanced.

References

Barnes, M. (2013). *Role reversal: Achieving uncommonly excellent results in the student-centered classroom*. ASCD.

Bernhardsson, L. (2023). Work-integrated learning through peer narratives about workplace experiences. *International Journal of Work-Integrated Learning, 24*(2), 277. https://www.ijwil.org/files/IJWIL_24_2_277_291.pdf

Black, P., & William, D. (2009). Developing the theory of formative assessment. *Educational Assessment, Evaluation and Accountability, 21*(1), 5–31. https://doi.org/10.1007/s11092-008-9068-5

Butterworth, C. (2024). *A reflection of authentic assessment design in two remote sensing case studies*. SaLIS.

Chirikov, I., Shmeleva, E., & Loyalka, P. (2020). The role of faculty in reducing academic dishonesty among engineering students. *Studies in Higher Education (Dorchester-on-Thames), 45*(12), 2464–2480. https://doi.org/10.1080/03075079.2019.1616169

Cobb, P. (1994, April). Where is the mind? Constructivist and socio-cultural perspectives on mathematical development. Annual meeting of the American Educational Research Association, New Orleans, LA.

Cook-Sather, A., Salmeron, D., & Smith, T. (2023). Humanizing STEM education through student-faculty pedagogical partnerships [Review]. *Frontiers in Education, 8*. https://doi.org/10.3389/feduc.2023.1153087

Council of Chief State School Officers. (2018). Revising the definition of formative assessment. https://ccsso.org/sites/default/files/2018-06/Revising%20the%20Definition%20of%20Formative%20Assessment.pdf.

Dewey, J. (2004). Democracy and education (Dover ed.). Dover Publications.

Drenth, P. J. D., Thierry, H., Wolff, C. J., & d. (1998). *Handbook of work and organizational psychology* (2nd ed.). Psychology Press.

Eaton, S. E. (2021). Plagiarism in higher education: Tackling tough topics in academic integrity, ABC-CLIO.

Eaton, S. E. (2022). Contract cheating in higher education: Global perspectives on theory, practice, and policy. *Palgrave Macmillan*. https://doi.org/10.1007/978-3-031-12680-2

Eaton, S. E. (2023). *Academic integrity in the age of artificial intelligence* [Keynote presentation]. Open Technology in Education, Society, and Scholarship Association (OTESSA), York University, ON. on May 29, 2023.

Eaton, S. E., Guglielmin, M., & Otoo, B. (Eds.). (2017). *Plagiarism: Moving from punitive to proactive approaches*. https://doi.org/10.11575/PRISM/31643.

Eaton, S. E., & Kumar, R. (2023). *Academic integrity lessons: Practical ideas for teaching, learning, and assessment.* University of Calgary. https://prism.ucalgary.ca/items/5c56dc44-5044-4fc4-baf1-9fa4b4c5470d
Fulford, A. (2021). Academic integrity and the disintegration of pedagogy. *Philosophy and Theory in Higher Education (Online), 3*(2), 23–42. https://doi.org/10.3726/ptihe022021.0002
Glowatz, M., & O'Brien, O. (2017). Academic engagement and technology: Revisiting the technological, pedagogical and content knowledge framework (TPACK) in higher education (HE): The academics' perspectives. *IAFOR Journal of Education, 5*(SI). https://doi.org/10.22492/ije.5.si.06
Hart, C., Hammer, S., Collins, P., & Chardon, T. (2011). The real deal: Using authentic assessment to promote student engagement in the first and second years of a regional law program. *Legal Education Review, 21*(1), 97–121. https://doi.org/10.53300/001c.6244
Higgs, J. (2014). Assessing the immeasurables of practice. *Asia-Pacific Journal of Coorperative Education, 15*(3), 253–267.
Hockings, C. (2010). *Inclusive learning and teaching in higher education: A synthesis of research.* Higher Education Academy.
Hodgkinson, T., Curtis, H., MacAlister, D., & Farrell, G. (2016). Student academic dishonesty: The potential for situational prevention. *Journal of Criminal Justice Education, 27*(1), 1–18. https://doi.org/10.1080/10511253.2015.1064982
Hummel, L., & Butterworth, C. (2024). *Case studies of authentic assessment in team activities in geomatics engineering technology*. SaLIS.
Janesick, V. J. (2006). *Authentic assessment primer.* Peter Lang.
Jones, L. (2007). The student-centered classroom. *Cambridge University Press.* https://mail.brettwilkin.com/phocadownload/StudentCentredClassroom/jones-student-centered.pdf
Khaled, N., & Lois, N. (2019). Student engagement in capstone projects through integration of technologies, project assessment and academic integrity. *Journal of Systemics, Cybernetics and Informatics, 17*(3), 26–30.
Martin, A. J. (2008). Enhancing student motivation and engagement: The effects of a multidimensional intervention. *Contemporary Educational Psychology, 33*(2), 239–269. https://doi.org/10.1016/j.cedpsych.2006.11.003
Martin, A. J. (2010). Building classroom success: Eliminating academic fear and failure. *Continuum.*
McCabe, D. L., Trevino, L. K., & Butterfield, K. D. (2001). Cheating in academic institutions: A decade of research. *Ethics & Behavior, 11*(3), 219–232. https://doi.org/10.1207/S15327019EB1103_2
Montiel, C. K., & Basil, B. S. (2024). *The amazing stories of point and cloud. MAPS362: 3D modelling*. SAIT.
Paull, M., Lloyd, N., Male, S. A., & Clerke, T. (2019). Engineering work integrated learning placements: The influence of capitals on students' access. *Journal of Higher Education Policy and Management, 41*(5), 534–549. https://doi.org/10.1080/1360080X.2019.1646382
Pecorari, D. (2015). Plagiarism in second language writing: Is it time to close the case? *Journal of Second Language Writing, 30*, 94–99. https://doi.org/10.1016/j.jslw.2015.08.003
Tai, J., Ajjawi, R., Bearman, M., Boud, D., Dawson, P., de St, J., & Jorre, T. (2023). Assessment for inclusion: Rethinking contemporary strategies in assessment design. *Higher Education Research and Development, 42*(2), 483–497. https://doi.org/10.1080/07294360.2022.2057451
U.S. Geological Survey. (2023). *EarthExplorer*. https://earthexplorer.usgs.gov.
von Glasersfeld, E. (1991). Introduction. In E. von Glasersfeld (Ed.), *Radical constructivism in mathematics education*. Kluwer Academic Publishers.
von Glasersfeld, E. (1995). *Radical constructivism: A way of knowing and learning*. Falmer Press.
Vygotsky, L. (1978). *Mind in society*. Harvard University Press.
Walshe, G. (2020). Radical constructivism—von Glasersfeld. In B. Akpan & T. J. Kennedy (Eds.), *Science education in theory and practice: An introductory guide to learning theory* (pp. 359–371). Springer. https://doi.org/10.1007/978-3-030-43620-9_24

Wiggins, G. (1989). A true test: Toward more authentic and equitable assessment. *Phi Delta Kappan, 70*(9), 703–713.

Carina Butterworth has worked as a geomatics engineering technology instructor at SAIT for the past 15 years. She is a professional engineer and holds a Master's Degree in Geomatics Engineering from the University of Calgary. She is currently a Ph.D. candidate at the University of Calgary in Biomedical Engineering. Her passion lies in teaching remote sensing and reality capture technologies, but she also enjoys teaching GIS and other software with field data collection courses. Carina is the faculty liaison for the Geomatics Engineering Technology Student Club, faculty representative on the SAIT Academic Faculty Association, faculty representative on SAIT's Academic Council, was the graduate student representative on the Biomedical Engineering Society's Education Committee, and was a Calgary chapter coordinator for the Canadian Association for Girl's in Science. She has been nominated 13 times for the SAITSA Teaching Excellence Award, was a recipient of the award in 2021, and has received numerous additional awards for leadership and collaboration efforts over the years.

Chapter 13
Exploring Collusion as a Form of Academic Misconduct in Vocational Shop Classrooms

Emily MacIsaac

Abstract Academic integrity literature from the last 30 years has focused primarily within the university context and on academic misconduct in written assignments, such as citation errors, copying and pasting, contract cheating (Eaton, *Plagiarism in higher education: Tackling tough topics in academic integrity,* 2021), and ghost-writing with new advancements in generative artificial intelligence (genAI). In general, the vocational college perspective of academic integrity has been neglected. Vocational programs typically provide students with hands-on skills training that is related to specific trades. Students of vocational programs acquire sufficient knowledge and skills in trade shop classrooms to gain employment on job sites immediately after graduation but also continue to build their skills in the trades. Trade shop classrooms are primarily designed to be open spaces where students can work together, but an important question that vocational program instructors face is where is the line between collaboration (e.g., sanctioned peer support) and collusion (e.g., academic misconduct)? This chapter delves into collusion in trade shop classrooms, investigating how instructors promote independent mastery in collaborative learning environments. To explore this issue, the author, a librarian, provides a summary of the literature on collusion, including its definition and potential manifestations in trade shop classrooms. Interviews with five trades instructors in Atlantic Canadian vocational colleges provide insight into the experiences of faculty and students in vocational education and offer tips for teaching practices that support academic integrity in trade shop classrooms.

Keywords Academic integrity · Unauthorized collaboration · Collusion · Vocational education · Vocational college · Shops and labs · Trade shop classroom · Industry training · Workplace integrity

E. MacIsaac (✉)
Holland College, Charlottetown, PE, Canada
e-mail: ermacisaac@hollandcollege.com

L. Vogt, B. M. Stoesz (eds.), *Academic Integrity in Vocational and Polytechnic Education*, Ethics and Integrity in Educational Contexts 8,
https://doi.org/10.1007/978-3-031-90091-4_13

Introduction

As a *Student Engagement and Instructional Librarian* at a small vocational college in Atlantic Canada, my primary role is to facilitate single in-class sessions on a variety of topics including, but not limited to, appropriate use of APA citation style and avoiding plagiarism. The library staff provides support to students, such as reviewing citations within written assignments to support students' appropriate implementation of APA documentation and formatting style, assist with scholarly research, and provide resources for other academic needs. Instructors also seek guidance from librarians when they encounter plagiarism within student assignments. Through these activities, I am steeped in conversations about plagiarism, leading to my interest in understanding why students plagiarize and what can be done to promote academic integrity within post-secondary. I typically support programs where lecture-based instruction with written assignments and exams are common. However, I have had opportunities to learn about assessments that require students to demonstrate their acquisition of skills associated with specific trades. Due to the necessity of collaboration in vocational shop classrooms, students may be more likely to engage in collusion than plagiarism. Unfortunately, much of the literature about academic misconduct describes the prevalence of these behaviours within the university context and there remains a lack of literature in vocational education. Consequently, this sparked my interest in when and how academic misconduct, specifically collusion, manifests in trade shop classrooms where students train for professional careers such as automotive or wind turbine technicians, welders, carpenters, electricians, and more. This chapter begins with a summary of literature describing collusion as a form of academic misconduct to understand the significance of this inappropriate behaviour in trade shop classrooms. To gain further insight, I interviewed instructors who teach trades and interact with students in shop environments in vocation programs in Atlantic Canada. Instructors shared several useful tips and tricks to reduce collusion and other forms of academic misconduct and promote academic integrity. This qualitative research study begins to fill the gap in the literature about academic misconduct.

Vocational Program Setting

Vocational colleges in Atlantic Canada offer trades programs (e.g., carpentry, plumbing, industrial electrician) similar to other vocational training institutes across Canada. Each student enters their program with a skill level different from that of their peers. After graduation, they have the skills to directly enter a job site where the techniques are applied. Additionally, students have the opportunity to earn a Red Seal Certificate after completing the appropriate level of blocks and/or working the required number of hours within a trade—although many tradespeople challenge their Red Seal after working the required number of hours (Program Excellence Red

Seal, 2016). Red Seal endorsement signifies that the knowledge and skills required for the Canadian national standard has been demonstrated by a tradesperson in their particular trade (Program Excellence Red Seal, 2022).

In the workplace, tradespeople often rely on each other to accomplish tasks and projects, but everyone typically completes specific work independently. Instructors of college trade programs aim to replicate industry conditions so students can more easily develop trade-specific skills in collaborative settings that are directly applicable to real-world job sites. To simulate the real-world environment, students spend more time in practical settings (e.g., training shop) learning specific vocational skills rather than focusing on understanding theory and writing assessments. There is a limited number of tools and specialized equipment in shop classrooms which often requires students to work in pairs or small groups, ensuring everyone gets hands-on experience to fully develop the necessary trade skills. For instance, a class might have only one table saw, which requires students to work in partners for training and safety purposes. Despite the benefits of collaborative learning, problems related to valid skill assessment may arise when students cross the line to collusion. My interviews with trade shop instructors examined how collaborative work in open teaching and learning spaces (e.g., vocational shops) impacts academic integrity, providing valuable insights into the strategies instructors use to promote positive academic behaviors in these environments.

Defining Collusion

In university settings, students may be accused of collusion when they have been "working with others without permission [or] is the product of unauthorized cooperation between the student and another person" (Sutherland-Smith, 2013, p. 51). Crook and Nixon (2019) expanded the definition of collusion by arguing that it occurs when students intentionally conceal working together on an assignment despite clear guidelines to complete the work individually. The distinction between collaboration and collusion may be unclear depending on the circumstances and assessment guidelines (Barrett & Cox, 2005; Sutherland-Smith, 2013; Sutton & Taylor, 2011). Indeed, Sutherland-Smith found that students in Australia were unsure of when collaboration crossed the line into collusion when completing assignments, and even "staff consider[ed] the term 'collusion,' as defined in the university policy, to be quite vague and difficult to put into action" (2013, p. 54). Vague academic misconduct policies within the Australian university context also resulted in students and faculty having differing opinions about helping versus academic misconduct (Sutherland-Smith, 2013). In some cases, students may use their own "moral code" to decide when to disregard 'individual work' assignment guidelines and ask for or provide help to their peers (Barrett & Cox, 2005). This may occur when students believe there are benefits to learning and that others are not significantly disadvantaged, particularly when permitted support is unavailable for challenging assessments (Barrett & Cox, 2005).

Understanding Collusion in Vocational Programs through Instructor Interviews

Given my past experience working in a vocational college library, collusion seemed the most obvious form of academic misconduct that could occur within a trade shop classroom. A student enrolled in a vocational program asking for help with a hands-on project may believe they are collaborating rather than seeking inappropriate assistance and are relying on their own ‘moral code’ to make this distinction. To learn more, I received multijurisdictional ethics approval from several colleges in Atlantic Canada, where I interviewed five instructors in early 2024 upon their written informed consent. The interview questions were designed to gain insight into how vocational students typically learn and work within the shop, whether behaviours classified as academic misconduct occur, and if trades instructors could share tips to prevent academic misconduct (see Appendix A for interview questions). All interviews took place virtually using Microsoft Teams and were recorded for data collection and review. To maintain anonymity, video files were deleted, and the names of participants and their institutions are not disclosed in this chapter.

Demographic Information

All participants identified as Caucasian. Four identified as men and one as a woman and one identified as being part of the 2SLGBTQIA+ community. All participants were between 40 and 55 years of age, with 5–40 years of experience working within their industry before beginning their teaching careers. Participants also indicated they had 3–20 years of teaching experience at their respective vocational colleges.

Themes Identified from the Interview Responses

Theme 1: Program and Shop Classroom Design

Most shop classrooms were described as large open spaces containing a variety of equipment, with student cohorts ranging from 10–21 students. Each instructor noted that they set up their classroom and learning activities in a way that simulated the real-world job site, and students consistently worked with partners.

- Every instructor worked within their trade before entering the teaching profession, so they know that tradespeople work as teams for most of their projects. With that being said, instructors also indicated that assessing the products submitted from individual learning activities is an important aspect of determining an overall grade that is representative of students’ learning.

- Instructors understand that each student enters the program with varying skill levels and physical capabilities. If a student lacks confidence due to insufficient trade knowledge or cannot physically complete a task (e.g., bending pipe), they may seek help when it is inappropriate to do so.

Theme 2: Struggling Students

Shop classes typically have one instructor and one lab assistant to support students, so it can be challenging to prevent inappropriate helping or collusion, especially with larger cohorts of students. In the open shop classroom, struggling students have more chances to seek assistance from stronger students.

- One instructor noted that students who struggle more than others will copy their peers' calculations or schematics when completing the homework that was assigned to prepare for a project within the shop.
- Instructors are also aware that skilled students may experience unbalanced workloads as they complete their own tasks and course assignments, all while supporting their peers. In such cases, instructors politely ask the skilled students to stop providing guidance. Two participants noted that most students respect these requests, allowing instructors to support the struggling students better.

Theme 3: Teaching Methods in Vocational Settings

Instructors encourage collaboration and peer support as it allows everyone to gain confidence in themselves as they learn new skills. Formative assessment was discussed by all participants to gauge how their students were progressing through core skills. For example, task sheets to grade how well a task was completed and how much teacher or peer assistance was required to complete projects. Allowing resubmissions to improve grades was another important component of several instructors' courses.

- An instructor provided insight into how a program cohort completes the same project in their shop classroom. The instructor first explains important elements from the blueprint and highlights common pitfalls before allowing students to attempt the task while overseeing progress. The entire class moves through project phases sequentially, with the instructor breaking down complex tasks into manageable parts (i.e., scaffolding).
- Instructors can discover when a student is not grasping a skill or that academic misconduct is happening by creating assessment checkpoints. For example, for students to move on to the next in-class project, they must describe and demonstrate to the instructor how they completed the first task. If students cannot explain their work, it raises concerns about whether learning has occurred.

Students can earn failing grades, but instructors often provide formative feedback, demonstrate the task, and then ask the student to repeat the assignment (i.e., pseudo-mastery learning model).
- Instructors mentioned that clear expectations in the daily functions within the shop are fundamental to their students' success. Setting boundaries before summative assessments ensures students understand that collaboration is not allowed in that context despite it otherwise being the classroom norm.
- The instructors understood the significance of peer support and collaboration to foster confidence and skill development. Instructors were also aware that most students acquire the necessary skills and feel self-assured in their abilities by the end of the term, so students need less help from peers and instructors by the time the final summative assessments are conducted.

Tips and Tricks for Trades Instructors

As the final question in the interview, I asked the participants to share tips that could help other instructors who may be struggling to prevent or reduce collusion in their shop settings. The following is a summary of the advice shared by the instructors.

Switch Partners

Instructors can switch student partners to assess student progress. If the original pair depends too much on each other, it limits skill development, and this dynamic will become noticeable. Switching partners also teaches students how to work with a variety of people which is applicable to industry.

Practice Independent Work

To prepare students for the final exam where collaboration is not allowed, instructors choose one or two classes during the term where students must complete pre-work (such as calculations or drawings) and classroom work independently. This strategy helps students hone their skills and allows instructors to identify those who need additional support before the final exam.

Reduce Stress in the Learning Environment

Students feel supported when instructors normalize struggles and encourage students to embrace mistakes. By reminding students that making mistakes is part of the learning process, students are encouraged to reach desired competencies by the end of the program. Offering opportunities to build confidence by completing large projects in a step-by-step format also helps students complete tasks independently.

Focus on Career Preparation

Discussing students' future careers in the specific trade can help students understand the severity of collusion or other types of academic misconduct. When students do not learn how to complete a skill during their time at college, they will not be prepared when they enter a job site. Instructors recommended creating an authentic environment that replicates a real-world job site as another way for students to prepare for the future in their trade.

Focus on Skill Development over Theory

When students have faced difficulties in secondary school, it can impact their confidence causing them to doubt their ability to succeed, especially during the theoretical or classroom portion of the vocational program. This is when an instructor can reframe a student's insecurities with learning by reminding them that core hands-on skills are the most valuable and that theory informs practice. A participant noted that if a student is anxious about literacy or numeracy, they are often pleased to hear that instructors value the ability to *do* rather than discuss.

Monitor Students Effectively

Stagger start times. If an instructor has a larger class, staggering start times can ensure that an instructor has eyes on their groups when they are at different stages of a project.

Small groups. Arranging a large cohort into teams of two or three makes it easier to monitor individual behaviour than groups of six or more.

Learning management system. Although not directly related to the shop classroom, one participant noted that data obtained from the learning management system (e.g., Moodle, Brightspace, or Blackboard) can indicate academic misconduct. If a student submitted an assignment but did not open the corresponding module in the learning management system, the instructor could be prompted to ask the student how they could have known what to do and how to do it.

Plan Assessments Carefully

Permitting students to resubmit. Students report reduced anxiety when they can continue to practice and then resubmit assessments to ensure they meet the skill requirement to receive a passing grade.

Scaffolding assignments. Creating assignments that build upon each other and require submission after completing each phase allows instructors to glean how their students develop their knowledge and skills at various stages of a course.

Allow opportunities for more practice. Shop time may be limited throughout the term based on a curriculum schedule that is intertwined with theory classes. When possible, and if a student is interested, instructors can offer materials for students to take home and practice independently. However, this strategy may not be appropriate for all trades programs, especially when safety considerations and/or specific equipment or tools exist.

Discussion

The interviews with vocational instructors revealed that they use a variety of teaching and learning strategies to encourage students to take ownership of their learning and become motivated to learn with integrity, including the following:

- **Student-centered learning** by providing students with a choice of a variety of authentic assessments, with clear guidance on classroom expectations and grading rubrics, allowing students to move at their own pace and focus on internal motivators (Starkey, 2019).
- **Scaffolding learning** is implemented in the shop classroom, and instructors encourage collaboration until they believe students have gained enough skills to work independently (Sotiriadou et al., 2020).
- **Mastery learning** is a key element in shop settings as the pacing of assignments is often flexible, and students can redo work and show instructors the skills they have obtained (Siddaiah-Subramanya et al., 2017).

Vocational instructors in this study use these methods in classrooms designed to mimic real-world contexts to motivate students, a strategy which can reduce academic misconduct, including collusion (Bertram Gallant, 2017; Sotiriadou et al., 2020). Creating and implementing authentic assessments with specific scenario-based contexts is important because "scenarios not only contextualize the assessment but enable students to see the relevance to real-world situations and reflect/link the theoretical aspects of theory course to industry practice" (Sotiriadou et al., 2020,

p. 2145). By adding future workplace authenticity to assessments, students may feel motivated to do well because they understand how the task fits into their future careers (Bertram Gallant, 2017).

Bertram Gallant (2017) discussed the importance of 'mastery-oriented environments' (p. 89) in teaching and learning contexts because this type of pedagogy makes students aware of the knowledge they already possess and motivates them to find a path to grow their skills. Deeper learning is more likely when students can resubmit an assignment or practice with an instructor, encouraging problem-solving or critical thinking (Bertram Gallant, 2017). Ultimately, when "instructors focus on asking 'what can I do to ensure learning?' rather than 'what can I do to reduce cheating?', they will discover new and invigorating approaches to creating classrooms in which cheating is the exception and academic integrity is the norm" (Bertram Gallant, 2017, p. 92).

Strengths and Limitations

The goal of this study was to broaden my perspective as a librarian in a vocational college and to allow trades instructors to advise others on best practices to support academic integrity. A strength of the study was using a qualitative research design to understand the experiences of instructors more fully.

One area from the interviews that remains unclear was whether academic misconduct occurred during the theory section or specific theory courses within shop-based programs. It was evident that students are required to understand and be tested on theory modules before entering the hands-on section of a program as well as take courses such as Computer Essentials. However, the interview questions focused on the shop component of the course, so we did not discuss academic misconduct within theory assessments.

Another limitation of this research was that the interview questions did not spark discussion about how collusion might affect the safety of future co-workers and clients due to insufficient industry knowledge. For instance, if students rely too heavily on classmates during the learning process, they may pose a danger on a job site. Future inquiry of vocational instructors' observations and perspectives about academic integrity and academic misconduct are needed to determine how instructors and students in vocational programs associate academic integrity with workplace safety.

Conclusion

Valuable insights and recommendations that can benefit trades instructors and support staff (e.g., librarians) in vocational education emerged from this project. The in-depth interviews from this qualitative study revealed that academic misconduct in the form of collusion is a concern for trades instructors who use a range of teaching and learning approaches to ensure that students learn and learn with integrity. Vocational students are often encouraged to collaborate because the nature of the training and real-world shop environments are inherently cooperative. Collaboration is also important for student growth as studying in vocational trades programs is a challenging process that requires hands-on practice and perseverance. When the challenges are too great, however, some students may find that they are seeking support from peers when inappropriate to do so. Vocational instructors, therefore, play a crucial role in creating supportive environments where students are encouraged to practice until they succeed and ask for support from instructors to build confidence in their developing skills. Working from this perspective offers a proactive approach to supporting academic integrity in vocational education.

Appendix A: Interview Questions

1. How do students usually work in the shop?
2. Do you think issues of academic misconduct (cheating) exist in shop settings?
3. If so, how often does cheating occur within the shop? For example, more than 50% of the time, less than 50% of the time, or almost never.
4. Have you ever experienced another student providing their peers 'too much' help in the shop, and was it ultimately detrimental to someone's learning?
5. How do you prevent academic misconduct from occurring in your shop?
6. How do you support students gaining core industry techniques/skills? (For example, hands-on skills to perform a certain task successfully)
7. What are your tips for other trades instructors struggling with cheating in their shop classrooms? (Such as discussing cheating at the beginning of the term and working with the students to ensure they know the skill or other instruction methods).

References

Barrett, R., & Cox, A. L. (2005). 'At least they're learning something': The hazy line between collaboration and collusion. *Assessment & Evaluation in Higher Education, 30*(2), 107–122. https://doi.org/10.1080/0260293042000264226

Bertram Gallant, T. (2017). Academic integrity as a teaching & learning issue: From theory to practice. *Theory Into Practice, 56*(2), 88–94. https://doi.org/10.1080/00405841.2017.1308173

Crook, C., & Nixon, E. (2019). The social anatomy of 'collusion'. *British Educational Research Journal, 45*(2), 388–406. https://doi.org/10.1002/berj.3504

Eaton, S. E. (2021). *Plagiarism in higher education: Tackling tough topics in academic integrity.* Bloomsbury Academic.

Program Excellence Red Seal. (2022, July 28). *Home.*. https://www.red-seal.ca/eng/w.2lc.4m.2.shtml

Program Excellence Red Seal. (2016, October 7). *Register for your red seal examination.* https://www.red-seal.ca/eng/resources/r.2g.3st.2r.shtml

Siddaiah-Subramanya, M., Smith, S., & Lonie, J. (2017). Mastery learning: How is it helpful? An analytical review. *Advances in Medical Education and Practice*, 269–275.

Starkey, L. (2019). Three dimensions of student-centred education: A framework for policy and practice. *Critical Studies in Education, 60*(3), 375–390. https://doi.org/10.1080/17508487.2017.1281829

Sutton, A., & Taylor, D. (2011). Confusion about collusion: Working together and academic integrity. *Assessment & Evaluation in Higher Education, 36*(7), 831–841. https://doi.org/10.1080/02602938.2010.488797

Sutherland-Smith, W. (2013). Crossing the line: Collusion or collaboration in university group work? *Australian Universities Review, 55*(1), 51–58. https://files.eric.ed.gov/fulltext/EJ1004398.pdf

Sotiriadou, P., Logan, D., Daly, A., & Guest, R. (2020). The role of authentic assessment to preserve academic integrity and promote skill development and employability. *Studies in Higher Education, 45*(11), 2132–2148.

Emily MacIsaac is the Student Engagement and Instructional Librarian at Holland College, a community college on the Canadian East Coast. Emily loves facilitating Library sessions on various topics, including APA and Plagiarism, to students in various departments at Holland College. Before working at the College, she received her Bachelor of Arts in History from the University of Prince Edward Island and her Master of Library and Information Studies from Dalhousie University. Emily has taken opportunities to write articles for edited publications throughout her academic career, such as the student-led UPEI Arts Review, Female Agency and Opportunity in the Italian Renaissance: Three Case Studies, as well as the Dalhousie Journal for Interdisciplinary Management, AIDS Represented within the Media, where she received a prize for her writing. Academic integrity is at the forefront of Emily's mind, so she is excited to learn more and write about non-written types of academic misconduct that can occur in College Trades shops.

Chapter 14
Automation, Creativity, and Ethics: Re-framing Academic Integrity in the Training of Music Technology Professionals

Hazel Fairbairn and Janette Tilley

Abstract College programs in Music Technology and Audio Engineering prepare students with a wide variety of creative and technical skills to meet the needs of a dynamic industry. While technical skills training is an important central piece of such programs, students are also introduced to the ethical and legal frameworks that guide and govern work in an increasingly globalized digital world.

Traditional academic integrity policies and discourses are often inadequate when considering the complex and often contradictory legal landscape that these students will experience. Moreover, recent developments in automated content recognition, as well as content generation itself, have radically changed what it means to do creative work and claim ownership over musical products and digital assets. This chapter explores emerging global issues in the college-level training of digital audio creators, producers, and engineers in British Columbia, Canada. While the specific legal requirements to which songwriters and musicians must attend have arisen from an increasingly competitive and litigious field, the principles of negotiated ownership and rights management may have wider application across disciplines as industries and workforce training increasingly value collaborative work. With text generation and automated content authorship now emerging as a concern for those working with the written word, the sonic arts may offer some lessons, as well as sobering warnings, about the future of automation in a world of human creativity.

Keywords Music technology · Copyright · Collaborative work · Music industry · Metadata · Song splits · Applied ethics

H. Fairbairn · J. Tilley (✉)
Douglas College, New Westminster, BC, Canada
e-mail: fairbairnh@douglascollege.ca; tilleyj@douglascollege.ca

L. Vogt, B. M. Stoesz (eds.), *Academic Integrity in Vocational and Polytechnic Education*, Ethics and Integrity in Educational Contexts 8,
https://doi.org/10.1007/978-3-031-90091-4_14

In her landmark volume on plagiarism in higher education, Eaton (2021) identified lacuna in the current academic literature, particularly with respect to the way in which non-text-based disciplines intersect with academic integrity policies. Eaton states that a "twenty-first century understanding of plagiarism" (p. 23) would include a great variety of artifacts and products from ideas to art, design, and music, yet the scholarly literature has largely remained mute. Simon, in Bretag's *Handbook of Academic Integrity,* devotes just four sentences to music (Simon, 2016, pp. 777–778) but concludes in general that the creative disciplines themselves "would need to form some sort of agreement as to what practices are appropriate in that discipline, what would constitute breaches of academic integrity, and how and in what circumstances references should be provided to external source material" (p. 779). Those of us working and teaching in the creative and technical arts often have an intuitive sense of what is appropriate based on professional practice, along with ethical and legal frameworks that shape our work. As we shift our focus away from plagiarism toward integrity more broadly, we invite the academic and vocational disciplines to articulate how ethical norms are practiced and honoured. Instead of a lacuna, we find a rich well of practices within music and the training of technical music professionals that have remained outside the discourse and research on academic integrity. A little creative thinking allows us to bring some of our experiences into the broader conversation.

Preparing Students for the Music Industry: Academic Integrity Meets Professional Ethics

Since being established as a 1-year certificate in 2011 and a 2-year diploma in 2016, the music technology program at Douglas College has built a strong reputation for nurturing students' creativity as musicians and creators and providing foundational technical skills in audio recording, music production, sound design, and live sound. The program also offers training in emerging specializations, including game audio, immersive sound, and postproduction. Most of our diploma graduates go on to work in the live event sector initially, which is where most entry-level opportunities exist. Many then build careers in the film industry (location sound and audio postproduction), game audio and/or studio engineering. The work is typically freelance, and successful professionals rely on their good reputation to maintain steady work in the field.

As musician-educator/scholar-practitioners, we have long been thinking about ways creative and academic practices intersect. Music, in general, is a live, competency-based discipline, and we both have experience with the highly competitive conservatory music education model as well as the more academic research-based scholarship of music. Audio engineering and music production are young disciplines within post-secondary institutions and have evolved in a pedagogical environment that aims to circumvent students' temptation to violate traditional academic integrity policies. Within our audio program's very practical training, there is a distinct effort to "keep things real." That is, focus on practical, real-world skills that apply rather than analyze concepts.

As a result, learning in the audio engineering and music production classroom tends to be assessed through low-stakes, process-focused practical assignments that consolidate rather than test learning. These are often completed during class time and, hence, under instructor supervision, where direct observational assessment is possible. Where an assignment is set within a Digital Audio Workstation (DAW) application, the submission criteria usually require the whole project folder to be submitted. If an auto backup feature is enabled, version histories are saved in a subfolder, making it relatively easy to detect inappropriate shortcuts in the workflow. There are minimal writing assignments, and these tend to be short and self-reflective tasks that carry very little weight or are part of industry-adjacent coursework such as professional writing and marketing. To develop the skills that a music producer or engineer needs, the assigned hands-on practical work is typically driven by intrinsic motivation to master a skill. Students practice with lower-stakes assignments that instill self-efficacy, an approach Lang (2013) identifies as particularly effective at dissuading cheating. While the end product is undoubtedly important to artists and listeners alike, the process of musical creation is iterative, and it affords opportunities for students to make the process visible to instructors, thereby dissuading cheating. Indeed, in our experience at Douglas College, very few academic integrity policy violations have been reported from the program.

While this may seem to invite complacency, do the low reports of academic integrity policy violations mean that we are really preparing students for the challenges of being professional creatives in a world mediated by algorithmic detection and governed by archaic copyright laws? Or is it simply a disconnect between integrity viewed through a traditional academic lens and the complexities of practical, professional ethics? It takes well-informed and mature critical thinking to navigate the professional music world, manage aggressive industry rights holders, understand the complexity of intellectual property law, and keep up with the fast-moving development of digital technologies.

Below, we consider three circumstances in which audio engineering and musical production students are confronted with professional ethical and legal norms as part of their training and how these intersect with the broader post-plagiarism concepts of academic integrity.

(Hip Hop) Values: Integrity and the Legal Landscape

Samples, loops, and remixes are core to modern music creation and, to an extent, shape contemporary views of ownership and creativity (Evans-Torkaryk, 2014; Lessig, 2008). Since hip-hop emerged in the late 1970s, strict rules of borrowing and citing have evolved to the extent that hip-hop sampling culture has been proposed as a model for teaching academic integrity in law classes (Chanbonpin, 2012).

Indeed, rather than fearing the influence of others, hip-hop and contemporary music, more generally, tend to lean into these borrowings, samplings, and auditory nods with the knowing listener invited to revel in the deep arcana of references (Lethem, 2007; Oswald, 1986). Mimic another artist's style too closely, and an artist risks the wrath of a protective fandom that demands originality and authenticity. The ethics of the community pass judgment quickly on those deemed to have appropriated another's art. Borrow actual sounds—loops, beats, samples—without permission or credit, and another moral and legal authority risks being violated—intellectual property and copyright law.

These well-evolved standards of integrity within hip-hop and remix cultures do not protect industry rights holders in the same way as writers and artists. In 2004, EMI lost a landmark cease and desist case against Brian Burton, AKA DJ Danger Mouse, who had created a critically acclaimed remix album from The Beatles *White Album* and Jay-Z's *Black Album*. The album had been created with the permission of the artists, in this case, Jay-Z (Shawn Corey Carter) and the two surviving Beatles, Paul McCartney and Ringo Starr. This set a precedent in the remix community that the approval of the original artists provides implicit permission for sample-based work and remixes (Rimmer, 2005). Of course, publishers and record labels do not see it the same way and dedicate entire departments to detecting copyright violations. Recent high-profile cases such as those around Ed Sheeran or Led Zeppelin have brought songwriting ownership to the attention of the wider public (Seabrook, 2023; Sisario, 2020). As industry rights holders have increasingly sought to protect their property, machine learning detection tools have been leveraged to identify similarities between songs, first looking at melodic similarity (Müllensiefen & Pendzich, 2009) and more recently at the waveform level. This manner of detection is akin to text-matching software, with algorithms locating matches within the music's fundamental sonic blocks. Emergent issues around the digital building blocks of popular music—samples—have thus far avoided the wider public gaze and complicated legal paradigms of copyright and moral rights.

Translating Academic Reference Practices to Creative Work 1: Metadata

The professional implications of this uneasy intersection between community-accepted norms, aggressive industry rights holders, and blind algorithms challenge the way that instructors teach music production. How do we reframe expectations around borrowing, re-using, re-framing, and re-mixing material in a way that prepares students for the realities of the professional world? Students aspiring to build a career creating musical and audio assets will need to learn to triage between artistic, legal, and professional demands and interrogate their own practice according to context.

Whether the musical material is human- or computer-generated, it takes many people to complete a music recording project. While the film industry has normalized the visibility of content creators, editors, and labourers via lengthy credit reels, such has not typically been the case in the music industry. With the demise of physical media (e.g., records, cassette tapes, or CDs) and liner notes, the long list of creative credit that any given audio track includes has been relegated to metadata. To the average consumer, metadata may seem, at best, inscrutable arcana or, at worst, irrelevant. The myth of solo creative genius governs the perception of musical objects; performers, rather than collaborators, figure overly large in the listener's imagination.

In an academic sense, metadata is the bibliography for a sonic asset. From a practical legal perspective, metadata is the mechanism by which copyright and credit is managed in a digital world. Giving credit appropriately is crucial for rights holders to be paid royalties. For students, the clash between ideas of solo creative genius and realities of documenting collaboration and borrowed material, provide a sobering learning opportunity and an invitation to practice industry-specific skills of integrity that will serve them well in the professional world. Appendix A offers a song metadata template for students to gather and document attributions for a track. Just as student writers need to take care to document all sources of information in their papers, so, too, do those working with multi-media projects. As our students across disciplines are increasingly producing digital works, from podcasts to short films, we find that prompting them to assign credit for all material used in a project aligns well with principles of academic integrity.

An audio assignment could require that, along with preparing the typical metadata for a release (title, genre), all writers, contributors, performers, session musicians, producers, and recording and mix engineers be credited. If the song uses samples, the minefield of rights clearance becomes evident as they track the progeny of a loop or hook. Many instructors set "sound alike" assignments, where students are challenged to reconstruct an existing track. This combines a robust framework for assessment with the development of students' critical listening skills. An addendum to this kind of assignment would explore the criteria for legal distribution by acknowledging the original rights holders in a way that might be akin to a well-cited essay. Furthermore, integrating such metadata collection into assignments would alert students to the world of intellectual property law in music.

Translating Academic Reference Practices to Creative Work 2: Song Splits

In the music industry, rights are negotiated and articulated via a song split contract, or "split sheet" (Gorman, 2021). A song split contract could be compared to a group work collaboration agreement, a pedagogical tool which guides groups toward equitable and honest work and holds members accountable for their roles (Brannen et al., 2021). A standard song split agreement requires all parties involved to negotiate two sets of copyright: the publishing, which traditionally deals with the melody and lyrics of the song, and the mechanical copyright, which is associated with the recording (see Appendix B). The latter would traditionally reside with the record label, who put up the money for recording, but it increasingly falls to artists as they fund their own creations without support.

Most artists who report having their songs stolen, recorded without permission, or claimed by someone else have had their copyright violated by a bandmate, collaborator, or friend, not by a malevolent industry "suit." It is common for young co-writers to have different notions of who owns the song, and some of these violations may be unintentional. Nonetheless, it is one of the most prevalent copyright infractions. Educating students to interrogate rights shares of creative collaborations while stakes are low protects and encourages them to engage with the notion of intellectual property in a relevant context.

A recent assignment in a music business course further demonstrates the power of this sort of industry-specific learning of integrity norms. The students at our institution recently collaborated in groups with students from a partner school in another country on another continent to create an original track and prepare it for release. The students in our institution recorded starting points (variously songs, beats, or bed tracks) and sent the building blocks to students in the partner institution to build upon. The assignment required students to send recordings back and forth, building the final product in layers. This workflow is entirely in line with industry norms as collaborative partners find themselves working asynchronously and over great distances. As the work developed, students negotiated their final song splits by assigning credit percentages. Interestingly, most chose to credit all participants equally, regardless of input.

The collaboration went well for most; however, in the confusion of sending files back and forth, two of the groups in the partner school accidentally used the same starting file containing a song written by one of our students. In this case, the song split agreement brought to light what might otherwise have been treated as an academic integrity issue: whether a mistake or a shortcut, the second group had not created an original song according to the assignment brief but had remixed someone else's. The requirement of the song split agreement in the assignment's workflow effectively caused the group to stop and think about their work and they reached out to the student creator on whose work their song was based. The student who wrote the original song was surprised to find how much she cared about being acknowledged in the contract prepared by the group who had appropriated it. The incident enriched the learning experience, and once both songs were contracted and credited appropriately, they did not violate copyright or academic integrity norms.

This example demonstrates how assessment in the audio classroom is very often process-driven group work, structured along the lines of professional teams. Students tend not to cheat because they are aware that they are building peer networks and the foundations of a professional reputation. On the rare occasion when academic integrity is violated, teammates are very quick to call out those who are not pulling their weight. Matching the internal motivations of teamwork with the external legal frameworks of copyright and digital rights management in the form of a song split agreement marries academic integrity and professional ethics in a visible way.

Future Challenges

Familiarity and collage, as opposed to wholescale novel invention, have long been integral to contemporary music creation, and license-free libraries of loops and sonic building blocks have been available for over a decade to facilitate this process: Apple Loops since 2009, and Splice since 2013 to name the best-known platforms. These tools have democratized, to a certain extent, some of the more technical elements of an audio project, making it possible for less-resourced artists, including students, to create work of greater complexity out of these building blocks. An artist need not negotiate splits or royalties via metadata when the material is already royalty-free.

For students, the use of open and license-free loops is a legitimate tool for learning technical skills, not just an aesthetic choice or favoured technique for artistic creation. Pre-built blocks of sound allow the audio student to focus on acquiring specific technical skills without having to first create the fundamental sonic material. While our academic policies might recognize the value of learning via legally sourced material, the proliferation of corporate "detectors" has made it practically impossible for students to showcase their work in a public setting on the web. Wave matching software, while pitched as copyright infringement detectors, cannot discern the difference between a legally licensed sample, a license-free loop (Splice or Apple), and unlicensed use of third-party material (Lawrence-Williams, 2022; Turnbull, 2014). If the waveform is recognized and matched to one already in the database from another previously released and copywritten song, the algorithm assumes copyright violation, resulting in a copyright strike against the platform user. This effectively outlaws legitimate tools and places them out of bounds for professional commissions, where a copyright strike would be a devastating career blow. Audio production students are all too aware of the challenges associated with breaking into a highly competitive industry and securing that first all-important professional contract; they cannot afford to jeopardize it by creating something that will be taken down by an algorithmic copyright strike.

Perhaps it is not surprising, therefore, that there is increasing awareness amongst young emerging artists of the limitations of using license-free samples. Since algorithmic copyright strikes have made it almost impossible to promote a track containing unlicensed material, even if that material is open and used legally, underground creators today are distributing remixes on obsolete and analog storage mechanisms such as cassette tapes and minidiscs to avoid detection (Tapper & Ahmed, 2024).

Of course, rapid changes in generative artificial intelligence mean that we are face-to-face with the prospect of wholly original (though also highly derivative) sonic material that may be created on demand. Suno, Loudly, Soundraw, AIVA and any number of other music generators further complicate the ethical and legal space our creative students work in (Finlayson, 2023a, 2023b).

Conclusion

Reframing academic integrity as professional integrity with a focus on intellectual property rights prepares students to navigate a world of archaic copyright legislation mediated by emerging digital technologies that still lack nuance. This approach also offers a mechanism to protect students from future exploitation of collaborative work.

Most programs silo Copyright Law in a separate music business course, missing the opportunity to have students research and acknowledge rights holders in all their derivative and remixed work. Embedding credits for rights holders across the curriculum in much the way that academic integrity policies are referenced in every syllabus would establish the integral importance and complexity of intellectual property in the music and audio industries. This is something that we are aiming to implement within our own program.

We are already seeing artificial intelligence-driven functionality impact the audio workflow within DAW software (Dredge, 2024). This will transform many audio professions. As we move into unpredictable redefinitions of the creative process, establishing the importance of intellectual property and the provenance of derivative work seems ever more important. Encouraging students to interrogate software applications for their appropriateness and question the material on which they were trained has the potential to protect rights holders as we rely ever more on algorithmic suggestions based on the work of "anonymous" creative artists.

From our experiences, we offer the following questions for instructors of other vocational training programs, particularly those preparing students for creative professions. Can professional templates for ownership of creative work and intellectual property, be it a copyright agreement, patent application or application to register a design, be used in the classroom to frame academic integrity within relevant and essential workplace skills? Can we extend this principle beyond contractual agreements between human creators so that students recognize both the opportunities and threats embedded in the algorithmic mediation of original work?

It is our hope that researching licensing terms and acknowledging generative music tools may protect creators from opportunist litigation, which is why we are embedding these outcomes in the music technology curriculum at Douglas College. As generative artificial intelligence continues to grow in the written sphere, we can envision similar outcomes for auto-generated text as authors challenge the use of their work as training data to grow large language models.

Appendix A

Song Metadata Template

Song Title: [Title of the Song]
Primary Information:

- **ISRC (International Standard Recording Code):** [ISRC Code]
- **UPC (Universal Product Code):** [UPC Code]
- **Release Date:** [Release Date]
- **Genre:** [Genre]
- **Language:** [Language]**Contributing Artists:**

- **Primary Artist:** [Primary Artist Name]
- **Featured Artists:** [Featured Artist Names]**Songwriters and Lyricists:**

- **Composer(s):** [Composer Name(s)]
- **Lyricist(s):** [Lyricist Name(s)]**Producers and Engineers:**

- **Producer(s):** [Producer Name(s)]
- **Mixing Engineer(s):** [Mixing Engineer Name(s)]
- **Mastering Engineer(s):** [Mastering Engineer Name(s)]
- **Recording Engineer(s):** [Recording Engineer Name(s)]**Performers:**

- **Vocalist(s):** [Vocalist Name(s)]
- **Instrumentalist(s):** [Instrumentalist Name(s) and Instrument Played]
- **Background Vocalist(s):** [Background Vocalist Name(s)]**Publisher and PRO Information:**

- **Music Publisher(s):** [Publisher Name(s)]
- **Performing Rights Organization (PRO):** [PRO Name(s)]**Sampled Material:**

- **Sample Title 1:**
  - **Source:** [Original Song/Loop Name]
  - **Rights Holder:** [Rights Holder Name]
  - **License Type:** [License Type]
- **Sample Title 2:**
  - **Source:** [Original Song/Loop Name]
  - **Rights Holder:** [Rights Holder Name]
  - **License Type:** [License Type]
- **Sample Title 3:**
  - **Source:** [Original Song/Loop Name]
  - **Rights Holder:** [Rights Holder Name]
  - **License Type:** [License Type]**Technical Metadata:**

- **BPM (Beats Per Minute):** [BPM]
- **Key:** [Key]

- **Duration:** [Song Duration]**Rights and Credits:**

- **Copyright Holder:** [Copyright Holder Name]
- **Mechanical Rights Holder:** [Mechanical Rights Holder Name]
- **Publishing Rights Holder:** [Publishing Rights Holder Name]**Additional Credits:**

- **Cover Art Designer:** [Designer Name]
- **Photographer:** [Photographer Name]
- **Video Director:** [Video Director Name] (if applicable)
- **Studio Name:** [Studio Name]**Contact Information:**

- **Primary Contact:** [Contact Name]
- **Email:** [Email Address]
- **Phone:** [Phone Number]
- **Address:** [Contact Address]

Appendix B

Song Split Agreement

Song Title: [Title of the Song]

This agreement is made on [Date], between the undersigned contributing writers and producers, collectively referred to as the "Parties," for the purpose of establishing ownership percentages in the song titled "[Title of the Song]", hereafter referred to as the "Song".

1. Publishing Rights: Contributing Composers, Producers, and Lyricists

Name	Role	Share of Song (%)
Name 1	[Composer / Producer / Lyricist]	[%]
Name 2	[Composer / Producer / Lyricist]	[%]
Name 3	[Composer / Producer / Lyricist]	[%]
Name 4	[Composer / Producer / Lyricist]	[%]
		= 100%

2. **Mechanical Rights: Master Recording** The Parties agree that the mechanical rights in the Song are to be split according to the percentages stated below.

Name	Share of Song Master (%)
Name 1	[%]
Name 2	[%]
Name 3	[%]
Name 4	[%]
	= 100%

3. **Sampled Material:** The Parties acknowledge and agree to the use of the following sampled material in the Song:

Sample Title	Source	License Type	Rights Holder	Amount Used
[Sample Title 1]	[Original Song/ Loop]	[Creative Commons/ Open/PRO]	[Rights Holder Name]	[%]
[Sample Title 2]	[Original Song/ Loop]	[Creative Commons/ Open/PRO]	[Rights Holder Name]	[%]
[Sample Title 3]	[Original Song/ Loop]	[Creative Commons/ Open/PRO]	[Rights Holder Name]	[%]
[Sample Title 4]	[Original Song/ Loop]	[Creative Commons/ Open/PRO]	[Rights Holder Name]	[%]

4. The Parties agree to clear all necessary rights and obtain any required permissions for the sampled material listed above.
5. **Agreement:** This agreement constitutes the entire agreement between the Parties with respect to the subject matter hereof and supersedes all prior and contemporaneous understandings, agreements, representations, and warranties, both written and oral, with respect to such subject matter.
6. **Signatures:**
   By signing below, the Parties acknowledge that they have read and understood this agreement, and agree to the terms and conditions stated herein.

Legal Name	Signature	Date
Legal Name 1	____	[Date]
Legal Name 2	____	[Date]
Legal Name 3	____	[Date]
Legal Name 4	____	[Date]

Contact Information:
For any queries or further correspondence, please contact:

- Legal Name 1: [Email 1], [Phone/Address 1]
- Legal Name 2: [Email 2], [Phone/Address 2]
- Legal Name 3: [Email 3], [Phone/Address 3]
- Legal Name 4: [Email 4], [Phone/Address 4]

References

Brannen, S. F., Beauchamp, D., Cartwright, N. M., Liddle, D. M., Tishinsky, J. M., Newton, G., & Monk, J. M. (2021). Effectiveness of group work contracts to facilitate collaborative group learning and reduce anxiety in traditional face-to-face lecture and online distance education course formats. *International Journal for the Scholarship of Teaching and Learning, 15*(2). https://doi.org/10.20429/ijsotl.2021.150205

Chanbonpin, K. D. (2012). Legal writing, the remix: Plagiarism and hip hop ethics. *Mercer Law Review, 63*(4), 597–638. https://digitalcommons.law.mercer.edu/jour_mlr/vol63/iss2/4

Dredge, S. (2024, May 15). *Apple's Logic Pro software now has an "AI-driven backing band."* Musically.com. https://musically.com/2024/05/15/apples-logic-pro-software-now-has-an-ai-driven-backing-band/

Eaton, S. E. (2021). *Plagiarism in higher education: Tackling tough topics in academic integrity*. Libraries Unlimited.

Evans-Torkaryk, T. (2014). Academic integrity, remix culture, globalization: A Canadian case study of student and faculty perceptions of plagiarism. *Across the Disciplines, 11*(2) https://wac.colostate.edu/docs/atd/articles/evans-tokaryk2014.pdf

Finlayson, A. (2023a, May 30). *AI and music-making part 1: The state of play*. Ableton.com. https://www.ableton.com/en/blog/ai-and-music-making-the-state-of-play/

Finlayson, A. (2023b, June 6). *AI and music-making part 2: Tomorrow is the question*. https://www.ableton.com/en/blog/ai-and-music-making-part-two-tomorrow-is-the-question/

Gorman, M. (2021, September 2). *Sound advice: Legal considerations when co-writing a song*. Words and Music. https://www.socanmagazine.ca/sound-advice/sound-advice-legal-considerations-when-co-writing-a-song/

Lang, J. M. (2013). *Cheating lessons: Learning from academic dishonesty*. Harvard University Press.

Lawrence-Williams, S. (2022). Regulating automatic content recognition software in Canada. *Intellectual Property Journal, 43*(3), 317–340.

Lessig, L. (2008). *Remix: Making art and commerce thrive in the hybrid economy*. Penguin.

Lethem, J. (2007). Ecstasy of influence. *Harper's, 314*, 1881. https://harpers.org/archive/2007/02/the-ecstasy-of-influence/

Müllensiefen, D., & Pendzich, M. (2009). Court decisions on music plagiarism and the predictive value of similarity algorithms. *Musicae Scientiae, 13*(1), 257–295.

Oswald, J. (1986). Plunderphonics, or audio piracy as a compositional prerogative. *Musicworks, 34*. https://www.plunderphonics.com/xhtml/xplunder.html

Rimmer, M. (2005). The grey album: Copyright law and digital sampling. *Media International Australia, 114*, 40–54. https://doi.org/10.1177/1329878X0511400106

Seabrook, J. (2023, June 5). *The case for and against Ed Sheeran*. The New Yorker.

Simon. (2016). Academic integrity in non-text based disciplines. In T. Bretag (Ed.), *Handbook on academic integrity* (pp. 763–782). Springer.

Sisario, B. (2020, October 5). Led zeppelin wins long "stairway to heaven" copyright case. *The New York Times*. https://www.nytimes.com/2020/10/05/arts/music/stairway-to-heaven-led-zeppelin-lawsuit.html

Tapper, J., & Ahmed, A. (2024, April 27). The "boring phone": Stressed-out genZ ditch smartphones for dumbphones. *The Guardian*. https://www.theguardian.com/society/2024/apr/27/the-boring-phone-stressed-out-gen-z-ditch-smartphones-for-dumbphones

Turnbull, F. (2014). The morality of mash-ups: Moral rights and Canada's non-commercial user-generated content exception. *Intellectual Property Journal, 26*(2), 217–236.

Hazel Fairbairn is an instructor and program coordinator for the Music Technology programs at Douglas College. Hazel moved to British Columbia in 2012 following a 25-year career as a musician based in the United Kingdom, collaborating with producer Mark Russell on a fusion jazz/folk/beat project which toured festivals and venues throughout Europe and North America. She came into education via Musical Futures, an informal learning initiative that emerged in the UK in the early 2000s, employing musicians to work alongside music teachers in the classroom. Consequently, she approaches teaching as a facilitator, committed to experiential, project-based learning. Alongside coordinating and teaching music technology at Douglas College, Hazel continues to work with Mark Russell and collaborates with award-winning poet Kim Trainor, creating string-based electronic soundscapes, sound design and scores for Kim's poetry films. Her doctorate is in Ethnomusicology from the University of Cambridge.

Janette Tilley is Associate Dean in the Faculty of Language, Literature and Performing Arts at Douglas College in New Westminster, British Columbia. She earned a Ph.D. in musicology from the University of Toronto and worked as a faculty member and music department chair in the United States of America, where she taught students in a variety of undergraduate and graduate music programs, from BA to Ph.D., before returning to Canada. While her musicological research focuses on music, gender, and religious practices, Janette has a keen interest in digital musicology. She is the editor of the Web Library of Seventeenth-Century Music, an open-access collection of modern scholarly editions published by the Society for Seventeenth-Century Music. As an academic integrity professional, Janette leads communities of practice at Douglas College.

Chapter 15
When Technology Uncovers the Truth: Exploratory Meetings in Architectural Technology

Chris Urbanski and Lisa Vogt

Abstract With a 2-year diploma in architectural technology, students can start their careers in the architectural, engineering, and construction sectors, supporting a team of professionals in bringing a client's vision to life. While graduates work collaboratively on large projects, students must demonstrate their developed skills and abilities through individual assignments. Despite being informed that the software used in the program tracks each plotted point, unauthorized collaboration occurs every year as assignments and pressures accumulate. This chapter explores how architectural technology instructors can determine with certainty whether academic work is original and how instructors at Red River College Polytechnic in Winnipeg, Manitoba, Canada, handle the submission of duplicated work. The chapter argues that even when there is clear evidence of an academic integrity breach, an exploratory meeting approach that focuses on supporting student growth allows instructors to prioritize learning and strengthen the culture of academic integrity within their discipline.

Keywords Architectural technology · Academic integrity · Software training · Exploratory meeting · Integrity breach detection · Discovery conversation · Student success · Industry preparedness · Canada

Introduction

My name is Chris Urbanski, and I am an architectural technologist and an instructor at Red River College Polytechnic (RRC Polytech) in Winnipeg, Manitoba, Canada. Collaboration forms the cornerstone of architectural technology projects and plays a key role in the learning experiences I design to prepare students for their future

---

C. Urbanski · L. Vogt (✉)
Civil Engineering Technology, Red River College Polytechnic, Winnipeg, MB, Canada
e-mail: curbanski@rrc.ca; lvogt@rrc.ca

L. Vogt, B. M. Stoesz (eds.), *Academic Integrity in Vocational and Polytechnic Education*, Ethics and Integrity in Educational Contexts 8,
https://doi.org/10.1007/978-3-031-90091-4_15

careers. When my colleague, Lisa Vogt, asked me to document my approach to using exploratory meetings in potential academic integrity breaches, I recognized I had no prior experience writing about my teaching practice for publication. We decided to use a collaborative approach to this chapter, and I took a brave leap to try something new. From my work in industry and in teaching, I believe that collaboration has many benefits, including leveraging each contributor's unique knowledge, skills, and abilities.

Collaborative Writing Process

My role in authoring this chapter was to provide the primary narrative based on my practical experiences with teaching architectural technology and leading exploratory meetings. I focused on sharing authentic insights and reflections from my teaching practice, aiming to offer a detailed account of how academic integrity issues are addressed through these meetings. The collaborative process allowed me to refine and better articulate my story. Lisa brought her expertise in academic integrity and her knowledge of writing for publication to the project.

We began with a recorded interview where Lisa asked questions, and I provided answers. Lisa then transcribed the interview and organized the transcript into topic headings and paragraphs, paraphrasing or rephrasing where necessary to enhance the flow. I worked from this document to refine and further organize the text, ensuring it reflected my voice and experience. Lisa added additional details on the policy and procedure in our institution and drew my attention to several academic publications that she integrated to support my perspective. Finally, we read through the text together and made the final adjustments that would result in this chapter.

This collaborative approach allowed us to create a vignette written from my perspective, documenting how an exploratory meeting—a topic represented in academic literature—is applied in practice. I have found that examining and documenting my approach has not only refined my teaching skills but also deepened my empathy for students who continually take risks and try new things as they progress through their educational program.

Architectural Technology Instruction

Our instructional team trains students to be architectural technologists in a 2-year program; graduates will work with architectural firms, engineering firms, residential builders, and the wider construction sector. Architectural technologists are detail-focused and play a vital role in translating a conceptual design into effective building plans. They must speak up when they notice a potential error or discrepancy in projects to ensure public safety and fulfill contractual agreements. Although integrity is not explicitly taught as part of architectural technology training, demonstrating integrity in coursework supports a student in building a positive reputation and allows an instructor to give a good reference to future employers.

Students in the Architectural Technology program learn how to work with Building Information Modeling (BIM) and Computer-aided Design (CAD) software, primarily Autodesk Revit, a comprehensive 3D modelling software, and Autodesk AutoCAD, a 2D drawing software. To teach students to use these tools, learning takes a scaffolded approach, starting with learning the fundamentals of the software before moving on to more complex tasks. In the early phases, students are expected to meet a desired outcome based on the learning and instruction of skills that have been presented. For example, students will draw a series of lines at a precise size and location. We are not looking for design decisions or interpretations in this type of early work.

On certain projects, we provide students with drawings or building models, either from previous assessments or prepared by the instructor, since we do not have time for the students to draw or model an entire building project in a 16-week course, let alone a single assessment. For example, students would be provided with a structural concrete foundation and then required to model structural rebar, demonstrating the correct sizing, location, and quantity of the structural reinforcing within that concrete component. This is scaffolded to previous assessments and allows the instructor to assess the student's understanding of specific applications and how to use the software tools. As a result, the student submissions should all look the same. This demonstrates the accuracy and precision of the student's skill and their ability to perform specific software processes to meet the learning outcomes of the assessment.

The teaching team routinely discusses providing unique assessments so that not every student would receive the same project. However, the impact on the instructor workload would be unmanageable to create 25–30 unique assessments for the typical six required projects in a course. In the future, we may have the opportunity to build a repository of two or three options per assignment and scale them by skill level so that instructors can choose an appropriate assignment for the group, but this would still lead to recycling projects year to year.

Education from the Start

I recognize the challenge of having 25–30 students submitting the exact same project, so maintaining academic integrity and identifying academic integrity breaches is ongoing. The students study together in a cohort for 2 years, and fostering a sense of community and collaboration is essential for students' growth in preparation for industry. Therefore, I prioritize early conversations about why integrity is important, encouraging students to contribute to the discussion and form their own ideas. This starts a verbal contract between the instructor and classmates about integrity.

When students install the software on their computers, they are informed that it tracks user activity to support collaboration and accountability in industry. These tools are helpful when managing industry projects to know who is working, when they are working, the time spent on tasks, and what tasks were executed. In an educational environment, this data can also be used to identify instances of inappropriate collaboration. These functions are embedded in the software and cannot be turned off. For example, Revit software tracks who opened the file, which user worked on the file, and the exact time to the second that work was saved. I am transparent with students about how the software tracks user information and that sharing files between students has led to academic integrity breaches and should not be done. If a student uses another student's file, the data will report multiple users and raise concerns about a potential academic integrity breach.

Architectural technologists in the workforce are always sharing files and collaborating on work. So, we talk about how to maintain the integrity of their schoolwork while developing important collaboration skills. We talk about good collaboration in their polytechnic program, which includes solving problems and finding solutions together. Building on the established culture of trust and communication, students are encouraged to talk to me if they are behind and need an extension. I have yet to meet a student who entered the program with the intention to engage in misconduct from the start, but academic integrity breaches occur when there are competing responsibilities, an overwhelming workload, or a lack of time to complete assessments.

I am fortunate to work with students throughout the duration of their 2-year program, allowing me to model the behaviour I want to see in my students and set expectations and standards that will benefit them throughout their academic and professional careers. Building a respectful community is the foundation for discussion, collaboration, and problem-solving. The students want to learn to work together, and these conversations are opportunities to learn how to do this appropriately while in school. These conversations also provide an opportunity to talk about what would be different in the industry and how collaboration will happen in that context.

Identifying Concerns During the Marking Process

During the marking process, there are three observable features that commonly lead to concerns about academic integrity breaches. These are described to demonstrate how technology can uncover the truth in the field of architectural technology.

Plotting at the Exact Same Coordinates

Because I have created these assignments and modelled the projects myself, I know the work thoroughly. The software does not give an automatic warning that someone other than the student has completed work, but I know the parts and can focus on details to quickly identify potential issues. For example, I would not prescribe specific coordinates to place a piece of text or dimensions to label elements on a technical drawing. The student determines the location of these elements to demonstrate their capabilities. The line work on the drawing may be specific to a particular coordinate; however, the chances of two students plotting a piece of text at the exact same coordinates to four decimal places of a millimeter are incredibly unlikely due to the detailed graphing of the workspace. As an instructor gets more experienced, they can more easily identify these details.

Matching User History

In Autodesk Revit, after the instructor has assessed the work, they can generate a report that displays information about the computer used, the date, and the time to the second. Again, this feature is in place to help professionals keep track of projects in industry. This history tells a different story in an academic setting because students receive grades based on their demonstrated abilities to meet the performance indicators. The instructor sees the student's movements, with the saved time and information about who modified the file. The instructor can observe any instances where the file was saved on a different computer, so if multiple usernames appear, the instructor will investigate further. If the time stamps on two submissions are identical, this indicates that the files are duplicates since students cannot perform the same exact movements at the same exact time on the same computer. In addition, part of the assessment requires students to include their initials when naming the elements they create in Revit and AutoCAD. If students panic and take another student's file to submit as their own in response to their panic, they must change the initials saved with each created element to reduce the chance of being caught cheating; changing initials is time-consuming, which acts as a deterrent for misusing files. This process supports assessment outcomes while having the student take ownership of their work.

Linking Unique ID Numbers

Students may pass a file back and forth and perform a "save as" to attempt to resave this file under a new name. But all the data is saved. For instance, Revit assigns a unique ID number to every single element a user creates. For example, placing a door gets an ID, drawing a grid line receives an ID, and so on. You cannot change these numbers in any way; every user has their own unique set of numbers. Another check we can apply is taking the unique ID numbers from one file and searching for them in another file. These ID numbers can be searched singularly or in large batches. If there is no connection between the two files, the reply from the software states that the ID number is not found. However, if students have copied work from another file, all the original ID numbers would be imported into the copied work, and the instructor can identify the specific model elements that have been duplicated.

Caveats

Differing Software It is important to note that not all my assessments are within Revit. Other programs track user movements in different ways. So, while I might still gather some evidence that suggests a problem, the evidence may not be definitive for every assignment. For example, AutoCAD offers different data that also points to anomalies in the user's work. With the addition of an AutoLISP, we can identify the author of each element in the drawing, and AutoLISP immediately notifies us if there is more than one author for the drawing element. Further investigation is always needed to confirm suspicions, and that is why we would need to ask the students questions to understand what has occurred more fully.

Access to Hardware Students in our program are required to provide and maintain a laptop capable of running specific software, but this does not always mean that the student has one well-functioning computer that they can rely on throughout their program. I have observed a range of reasons why students may legitimately work on multiple computers, including instances where they own multiple devices, share devices with family members, do not have access to the best-performing computers, or experience technical problems and use a temporary replacement computer. Although our department offers tech support, these challenges are difficult for students who often have reduced financial capacity during their studies. By contrast, graduates will have access to better-quality technology and robust in-house tech support when they are employed as professionals in industry.

Responding to Concerns

At RRC Polytech, faculty are the first level of response to academic integrity concerns and are guided by the A17 Academic Integrity Policy (Red River College Polytechnic, 2022) and Academic Integrity Procedures guide (Red River College Polytechnic, n.d.). Policy revision in 2023 incorporated recommendations from the exemplary academic integrity policy toolkit (University of South Australia, n.d.) and the core elements of exemplary academic integrity policy (Bretag et al., 2011). The policy explicitly names the responsibilities of the institution, faculty, students, and staff in upholding the values of academic integrity and promoting an academic integrity culture, shifting the approach to a shared responsibility. The procedures guide (Red River College Polytechnic, n.d.) introduces the exploratory meeting as:

> a supportive conversation between the instructor and student to discuss the instructor's identification of a potential academic integrity breach. The intended outcomes of the exploratory meeting include (1) sharing information (2) understanding context (3) educating the student, if needed and (4) identifying next steps. The exploratory meeting is the first step in the process to share information about the concern, understand the student's perspective and gather further information to help determine whether an academic integrity breach has occurred.

Faculty are responsible for inviting the student to the exploratory meeting, and they may call upon their departmental Chair or the Academic Integrity Specialist to assist and advise as needed. The staff intranet page provides further details to guide faculty, advising that the conversation should be approached with honesty, curiosity, and respect. The meeting is intended to gather information and is not a venue for making decisions. To conclude the meeting, faculty are to thank the student for participating and advise them that the next steps will be communicated after the meeting. The Academic Integrity Procedures Guide (2022) provides full details on how faculty should proceed. Similar processes in other post-secondary institutions may be referred to as discovery interviews (Eaton, 2018) or courageous conversations (University of New South Wales, n.d.).

Although the exploratory meeting concept was a new addition to the policy, I had employed this practice before it became part of the institutional process, as I believe it is a respectful way to proceed in a professional training program. This rationale is supported by Murdoch and House (2023), who describe the positive impact staff can have when they avoid enforcing power dynamics and instead rely on their humanity when addressing concerns about academic integrity breaches. I also know that my students are novices who have entered my course with the intention to learn. Having a conversation with a student about an academic integrity breach is not an inconvenience; instead, it creates a teachable moment and ensures that learning is continuous (Bertram Gallant, 2017).

The Exploratory Meeting

When I detect a marker that suggests that the student did not complete the work themselves, I meet with the student. If multiple students are involved, I will speak to them separately. I start by asking them to explain their process for creating the project as best as they can remember. I ask questions such as when they started the project, how long they worked on it, and how many hours or days they spent, all examples of data that can be extracted from their project files. A student who has completed the work with integrity should be able to reasonably explain their workflow; however, if the student struggles to recall basic details about their process, it raises the level of concern about the file's authenticity. At this point, some students will acknowledge inappropriate collaboration, allowing us to carry the conversation to the next steps. In other cases, the student might falsely recount what happened. At this point, I would present the evidence and ask the student how they would explain the record captured by the software. In my experience, students often admit that the tracking does express what actually happened.

Even when there is clear evidence of an academic integrity breach on the file, I still treat my response as a conversation, not as an accusation. There are several reasons for doing this. First, I always need to confirm what the evidence suggests and have the student contribute to understanding what happened. This allows the student to speak for themselves and own up to a mistake without being forced. This is a respectful way to communicate with people and reflects how people are treated in industry. Second, I view all parts of my course as learning experiences, from the first day when we start to share our expectations until the end of the course. This doesn't change if students make poor choices. There is no benefit to the instructor, institution, or industry in pushing the student to feel greater embarrassment or regret than they already feel. Although academic cheating can be a sign of poor decision-making, I do not want students to ruin their academic career or employment prospects over one poor choice. Third, having a conversation with the student creates an opportunity to resolve the matter in the most positive way possible. We do not use the term 'restorative justice,' but I do think that students need opportunities to rebuild good working relationships with their classmates and instructors. They also need to have faith restored in themselves, so I focus on building relationships rather than punishing crimes.

If an exploratory meeting determines that an academic integrity breach did not occur, I have always found that having the conversation has value in reinforcing what we discussed in the course. Students should be aware that their work is being monitored and checked for integrity so that they are encouraged to know that integrity matters.

Conclusion

In our industry, we are constantly balancing time, cost, and quality. Architectural technologists play an essential role in making the client and architect's vision come to life while considering the balance of the three. The responsibility to prepare students to meet the technology expectations and standards is equally matched by the need for our graduates to work ethically and respectfully with their future colleagues, managers, and clients. I maintain strong ties with industry leaders, and my reputation as an educator in our community drives me to keep integrity at the center of my teaching.

Although much of the architectural technologist training is focused on technical skills, fostering holistic development in graduates benefits the workplace and reflects positively on our program and institution. Exploratory meetings ensure that students are treated with respect and guided toward the program's intended learning outcomes. Even in cases where technology has uncovered the truth, and the evidence is strong and indicative of an academic integrity breach, approaching the situation with compassion and curiosity supports the repair of relationships and resolution of issues required to move forward. We want the architectural technology diploma to signify that students not only have technical competency but also a commitment to ethical practice in the workplace, demonstrated through fair collaboration and honest representation of their skills.

As students graduate and transition to working as architectural technology professionals, they will be responsible for performing tasks on high-cost projects, ensuring that designs and technical solutions are accurate and credible while contributing to the planning of life safety and accessibility details and specifications. If a graduate has not acquired sufficient skills in their program, their inability to work to industry expectations can have serious consequences for the reputation of a firm and the architect. In such circumstances, we cannot cut corners as educators, and we cannot allow our students to cut corners. Although architectural technologists do not take an oath of practice and the liability falls with the architects and engineers, a duty to perform and execute the tasks to the best of our abilities, meeting the complex standards set out for each project, is paramount. We must always uphold the highest level of integrity to meet these expectations and fulfill our roles as educators, students, and professionals.

References

Bertram Gallant, T. (2017). Academic integrity as a teaching and learning issue: From theory to practice. *Theory Into Practice, 56*(2), 88–94. https://doi.org/10.1080/00405841.2017.1308173

Bretag, T., Mahmud, S., Wallace, M., Walker, R., James, C., Green, M., East, J., McGowan, U., & Patridge, L. (2011). Core elements of exemplary academic integrity policy in Australian higher education. *International Journal for Educational Integrity, 7*(2). https://doi.org/10.21913/IJEI.v7i2.759

Eaton, S. E. (2018, September). *How to lead a discovery interview about contract cheating.* Taylor Institute for Teaching and Learning. https://taylorinstitute.ucalgary.ca/resources/how-to-lead-discovery-interview-contract-cheating

Murdoch, K., & House, D. (2023). Courageous conversations: Approaching amnesty through honesty as reparations to a learning community. In S. E. Eaton (Ed.), *Second handbook of academic integrity* (pp. 1–13). Springer Nature. https://doi.org/10.1007/978-981-287-079-7_181-1

Red River College Polytechnic. (2022, June 17). *A17—Academic integrity policy.* https://www.rrc.ca/legal/policies/academic-integrity/

Red River College Polytechnic. (n.d.). *A17—Academic integrity procedures.* https://www.rrc.ca/legal/policies/academic-integrity/procedures/

University of South Australia. (n.d.). *Exemplary academic integrity project.* https://lo.unisa.edu.au/course/view.php?id=6751&sectionid=112510

University of New South Wales. (n.d.). *UNSW courageous conversations.* Office of Assurance and Integrity. https://www.unsw.edu.au/planning-assurance/conduct-integrity/conduct-unsw/unsw-courageous-conversations

Chris Urbanski is an Architectural Technologist and a faculty member in the School of Skilled Trades and Technologies at Red River College Polytechnic. He is the Program Lead and an instructor in the Architectural Technology Program, committed to delivering high-quality, multi-discipline technical training to meet the current and emerging needs of his students, industry partners, and community. Chris' teaching focuses on current and emergent technologies utilized in the building design and construction sectors. Recognizing the vital link between academic training and the standards, expectations, and the role technology plays in the Architectural, Engineering and Construction sectors, Chris is dedicated to empowering students and industry professionals to embrace digital ways of working to support project design and execution.

Lisa Vogt has dedicated over 20 years to higher education, beginning her career as an English as an Additional Language (EAL) teacher in Tokyo, Japan, before transitioning to Canadian higher education. Currently, she is an instructor at Red River College Polytechnic, where she teaches leadership and ethics-focused courses. Lisa has published and presented on academic integrity at local, national and international venues, sharing research findings and practitioner insights. The common thread in her work is creating positive and inclusive educational environments that enable students and staff to thrive. She holds a B.A. in Intercultural Studies, a certificate in Teaching English to Speakers of Other Languages (TESOL), and an M.A. in Higher Education Administration and Leadership.

Chapter 16
English Language Support Served with Integrity: Reflections from a Hospitality and Tourism Management Workshop Facilitator

Emilie Jackson

Abstract Students in vocational programs in Canada whose first or primary language differs from the language of instruction face the dual challenge of mastering new industry-specific skills while demonstrating competencies in an additional language. At Red River College Polytechnic in Winnipeg, Manitoba, Canada, the Academic Success Centre supports Hospitality and Tourism Management students in acquiring specialized vocabulary and building confidence with industry-specific communication strategies. This chapter offers insights from the Academic Support Specialist who facilitates workshops designed to prepare these students to serve lunch guests at Jane's, an upscale fine-dining restaurant on campus. Building students' language and cultural skills for vocational career success creates opportunities to demonstrate the six fundamental values of academic integrity-honesty, trust, fairness, respect, responsibility, and courage-as outlined by the International Center for Academic Integrity (2021). Reflecting on the six fundamental values can guide institutional members in refining their professional practice and fostering a culture of academic integrity that supports students in achieving their academic and professional goals.

Keywords Academic integrity · International students · English language learners · Language other than English · EAL · Vocational training · Hospitality · Culinary arts · Canada

---

E. Jackson (✉)
Red River College Polytechnic, Winnipeg, MB, Canada
e-mail: emjackson@rrc.ca

L. Vogt, B. M. Stoesz (eds.), *Academic Integrity in Vocational and Polytechnic Education*, Ethics and Integrity in Educational Contexts 8,
https://doi.org/10.1007/978-3-031-90091-4_16

Growing a Culture of Academic Integrity

Educators in Canadian polytechnics teach skills that prepare learners for vocational careers, often supporting students whose first or primary language differs from the language of instruction. Building students' language and cultural skills for vocational career success creates opportunities to demonstrate the six fundamental values of academic integrity (International Center for Academic Integrity [ICAI], 2021), which are honesty, trust, fairness, respect, responsibility, and courage. It is essential for educators to demonstrate and promote these values to grow a culture of academic integrity (East, 2010; ICAI, 2021; Morris, 2018).

I am an Academic Support Specialist in the Academic Success Centre at Red River College Polytechnic in Winnipeg, Manitoba, Canada. Our centre plays an important role in responding to student academic support needs in one-on-one sessions, small groups, and workshops. I do not hold a formally assigned role to teach academic integrity, nor do I have responsibilities to identify or resolve potential academic integrity breaches. Instead, my role is to respond to student needs, which can involve discussing academic integrity policies and how students can demonstrate integrity in their learning. Students often feel more comfortable sharing their concerns about how to meet academic integrity expectations with staff in the Academic Success Centre because we do not have a part in the disciplinary elements of the academic integrity procedure. As a staff member at RRC Polytech, I have reflected on my role in building a positive culture of academic integrity. In this chapter, I share my experience teaching the target language and cultural expectations associated with working in a hospitality setting and examine how the six fundamental values of academic integrity ground my work.

English Language Supports in Hospitality and Tourism Management

First-year students in the Hospitality and Tourism Management (HTM) program complete the Restaurant Service Practical course that requires them to serve lunch guests at Jane's. Jane's is an upscale restaurant in a historic bank building in downtown Winnipeg with ornate three-story high ceilings and a three-course menu designed and prepared by Culinary Arts program students. Jane's is considered a 'training ground' for HTM students (RRC Polytech, n.d.) who fill the roles of servers and hosts by welcoming and seating guests on arrival, taking and serving guests' lunch and dessert orders, as well as preparing and serving beverages. The Restaurant Service Practical course is an exciting opportunity for students to develop practical skills to work in the Canadian hospitality industry. However, this excitement is often matched with worry for students whose first language is not English, as their experience at Jane's might be their first time having extended conversations with English first-language speakers in a professional service setting. The Academic Success

Centre collaborates with HTM instructors to offer English language workshops for these students, recognizing the challenges they face and their need for language support to meet the language demands of lunch service at Jane's.

The Restaurant Service Practical Course

The Restaurant Service Practical course consists of 3 to 4 weeks of in-class preparation and 6 weeks of practical lunch service at Jane's. Instructors teach restaurant theory and the fundamentals of Jane's operations during the in-class portion before students begin in Jane's. With a significant amount of content to cover, instructors do not have time to provide focused English practice and review all the cultural expectations that may be unfamiliar to students who are new to Canada. Instructors introduce students to Jane's lunch menu items, including preparation techniques and ingredient lists. However, English language learners often require additional opportunities to review menu terminology and practice saying the terms out loud before interacting with lunch guests.

English Language Workshop Series

When I first took over facilitating the workshops, they seemed straightforward—teach the students the target phrases and vocabulary and give them opportunities to practice before they get to Jane's. However, I have learned that teaching the language required for a specific cultural context to adult learners from various cultures requires a more thoughtful approach, especially given that these students have industry experience. As students learn the target language and cultural expectations for verbal and nonverbal communication at Jane's, they often compare what they are learning to their previous work experiences, or their expectations of service based on their cultural understanding of communicating in restaurant or hospitality settings. The English Language workshop series provides the content, skills, and practice that the course instructors are unable to provide. The series usually consists of eight 50-min workshops scheduled during students' free periods. The goal is for students to learn and practice the target language required to provide lunch service at Jane's so that they are ready when their rotation in Jane's begins, approximately 4 weeks after the start of each term.

Students must learn menu items that often include unfamiliar English terms and words from other languages, such as French, German, and Italian. In the workshops, students practice saying the words out loud, receive feedback on their clarity and pronunciation, practice explaining the menu items, and answer questions that lunch guests might ask while deciding what to order. A few examples of menu items from the winter 2024 lunch menu include:

- Bacon Wrapped Pork Tenderloin with Caramelized Onion Polenta, Grilled Broccolini, Candied Fennel and Jus Lié
- Blueberry Tart with Chantilly Cream and Blueberry Ginger Coulis

The lunch service steps include greeting the guest, taking the lunch order, serving the lunch order, taking and serving the dessert order, processing the payment, and thanking the guest. While students memorize the steps and practice the associated phrases individually, the workshops provide an opportunity to practice interactions and build confidence with the language, including responses to guest questions.

Students' Intersectional Needs and Identities

While working with students who speak English as an additional language, including international students, permanent residents, refugees, and new Canadians, I have observed these students experience a range of challenges due to language barriers, cultural differences, and institutional expectations. These challenges can be amplified or further complicated by their intersectional needs and identities, such as educational background, professional experiences, age, gender, and sex, as well as familial and financial pressures and responsibilities. During the first English language workshop, I always ask students why they selected the HTM program, and students often refer to past work experience in tourism or customer service industries and their desire to help and meet new people as their primary motivations for selecting a program to prepare them to work in the hospitality industry. However, while students may be enthusiastic about the opportunities to work in this industry in Canada, they often have concerns about how their English language skills and understanding of cultural expectations will factor into their ability to do the work and get a job in the hospitality and tourism fields.

Cultural Aspects of Language

Providing remedial support for this student population requires a thoughtful approach that acknowledges differences in culture. As the facilitator of the English language workshops for the HTM program, I have found it challenging to balance the needs of students and the expectations of the program while ensuring that "Western" cultural views and norms are not given priority over other cultures. The target language I teach in the workshops opens the door for many discussions around language and culture. Some of the most common topics of these discussions during the HTM English language workshops include ways of demonstrating respect and professionalism through word choice, tone, body language, and other nonverbal communication cues, as well as how to balance professionalism and friendliness in a service setting, and more granularly, how to refer to customers, request information, manage miscommunication, and more. During these discussions, my concern is that if I am not culturally sensitive, students may feel that their experiences and cultural values are less than the Western values they are being asked to adapt to and that their prior knowledge and experiences do not have a place in this new cultural setting. This type of exclusion can make students feel embarrassed, frustrated, discouraged, and demotivated.

The Fundamental Values of Academic Integrity

As I worked on preparing and perfecting these value-focused discussions with students, words such as respect, trust, and responsibility continually came to mind. These reflections made me realize that my approach to teaching the target language and engaging with students' questions about cultural expectations is often grounded in the fundamental values of academic integrity. While I have not been explicitly tasked to teach academic integrity during the workshops, the values are present in my teaching practice. The remaining discussion will reflect on how the values are demonstrated in the English language workshops to enhance students' academic success.

Honesty

As the workshop facilitator, I am responsible for being honest with the students about the difficulties they may face at Jane's while ensuring that students gain the confidence they need to succeed. Serving guests in a busy restaurant environment can be intimidating for English language learners. Throughout the lunch service, students must request a significant amount of information from guests and be prepared to respond to queries in a large space full of chatting diners. The noise level can make spoken interactions more challenging for students developing their English language listening and speaking skills. Furthermore, students regularly interact with guests who may not employ the same communication strategies used by their instructors and tutors to support English language learners.

I must be honest with students about their strengths and areas that need improvement so that students learn to speak over the significant noise level and communicate with unpredictable guests. During the first workshop, I talk about my expectations for workshop participation. I remind students that these are speaking workshops, so participating in the workshops means saying the vocabulary words and practicing phrases out loud in front of the class and with peers. I inform students that I will provide feedback on their speaking and ask the students to apply the feedback in the moment. When students practice with a partner, I also encourage them to provide their peers with feedback. By deliberately building honest feedback into every workshop activity, each student receives substantial input from me and their peers. As students receive feedback at each workshop, they can get a realistic understanding of what they need to work on and how prepared they are to interact with lunch guests at Jane's. When students have a clear understanding of their level and progress related to a specific goal, I have observed that this can increase student motivation to continue working towards mastery of that goal (Ambrose et al., 2010). Being honest with students can help students be honest with themselves (ICAI, 2021).

Trust

One of my goals as the facilitator is to build trust with students so that they can ask questions freely. I do this by taking time to answer their questions and giving them time to share their experiences. I have found that students seek support for multiple reasons, including challenges in their classroom environment, low confidence in post-secondary educational settings, and other barriers impacting their learning. English language learners often fear criticism from their instructors and peers based on their ability to communicate clearly in English (Kubota et al., 2023). By prioritizing building trust with students, students feel more comfortable taking risks with the new vocabulary and become confident that their classmates will support them in their learning journey.

For example, English language learners often have trouble pronouncing English words because the mouth shapes they need to make to create the English word sounds are different from the shapes they are used to making when speaking their primary languages. Practicing moving the tongue and lips in different directions can feel uncomfortable and embarrassing. Nonetheless, for students to succeed in pronouncing menu items, including unfamiliar French, German, and Italian food names, they need to risk feeling uncomfortable and getting it wrong to achieve the correct sounds. In a trusting environment, I have observed that students have reduced fear of embarrassment and can learn the target vocabulary required to become proficient servers at Jane's.

Fairness

As the workshop facilitator, my goal is for every student who consistently attends to have a fair opportunity to build confidence with the language needed to serve lunch guests at Jane's. Workshop participants have varying levels of English proficiency. Some students need extra support to learn the target language, so I make time for individual feedback during the workshops when possible. For example, during the second half of each workshop, students are put in groups of two or three to practice the language introduced during the first half. I circulate to answer questions and provide feedback, checking in with each student and giving targeted individual feedback if there are specific phrases, words, or sounds that a student needs to develop.

Conversely, a few students often have strong English language speaking and listening skills, so they pick up the target language quickly but continue to attend the workshops because they are motivated to take any opportunity to practice speaking English. There have been cases when these students dominated the workshops by answering every question I posed or asking questions about language that we had yet to cover. This creates an unfair situation for lower-level language students, who may become confused and feel insufficiently supported. While I answer the questions posed by confident English language learners, I must maintain a pace that does not cause the students with the most significant support needs to fall behind. In these cases, I remind all students of our workshop goals for that day and ask that we focus on those skills. To maintain a fair and encouraging learning environment where remedial skill-building is the basis of the workshop goals, I believe it is essential to prioritize the highest support needs. Classroom instructors cannot constantly adjust their pace to suit the needs of the students, so remedial workshops must prioritize the element of fairness to offer adaptability and support to the participants who struggle the most.

Respect

In a multicultural group of students with diverse experiences, it is important to ensure that each student's input is respected and valued. Workshop participants who have worked in restaurants or hospitality settings in other countries often have questions about why a particular language or vocabulary is being recommended or not recommended because it is different from what they have experienced in the past. They may feel that their past experiences are not respected or valued now that they are being asked to speak and behave differently. When students inevitably use vocabulary or phrases that have different implications in specific cultural contexts, I take the opportunity to address the entire class and open the floor for a discussion about language and respect in different cultures. During these discussions, I encourage all students to reflect on and share how they use language to show respect. In some cases, I share that some of those terms may be received differently due to cultural expectations and norms in the Canadian context. My goal for these discussions is that each student leaves the session understanding that while they may need to adapt to the expectations at Jane's, their cultural ways of showing respect through language are not wrong, and their experience and knowledge are valuable. Students are receptive to these discussions because they want to learn how to "get it right," but I also want students to consider adapting what they know instead of devaluing what they know.

An example of this is using "ma'am" to address feminine-presenting guests. Many students from different cultures use this term to show respect to women, so students use it when they practice greeting a guest at Jane's. However, some

Canadians may feel that this term is only fitting for an older generation, is too formal, or is poorly suited to the Canadian context. In addition, this term is associated with women, so using "ma'am" assumes the guest's gender. When discussing these topics, I show students respect by asking them about their cultures and ensuring that all cultures in the room can be represented if students want to share. This ensures that students feel their own culture is respected and provides an understanding of the Canadian cultural expectations necessary to succeed in the fine-dining setting at Jane's.

Responsibility

Students in the HTM program have many responsibilities, vying for their attention. Besides attending and completing coursework, many students work during evenings and weekends and have family responsibilities. Students attend six or seven classes each term, and additional time is scheduled to complete the certifications required to work in the hospitality industry in Manitoba. For students to willingly attend an additional eight hours of workshops, the English language lessons and activities must align with students' goals. If students feel that what they are learning will help them succeed, they will feel motivated to keep attending and learning, as motivation directly influences student effort and priorities in their learning (Ambrose et al., 2010). As the workshop facilitator, I have the responsibility of ensuring that the workshop content can be directly applied to practice at Jane's. This is achieved by collaborating with course instructors to ensure consistency between the learning objectives of the course and language workshops and to stay up to date with any changes to Jane's lunch service. This collaboration illustrates my responsibility to place student goals at the forefront of the workshop series design.

However, the responsibility is not entirely on me as the facilitator. The workshops are optional, so it puts the responsibility on the students to be proactive and take the opportunity to attend the workshops and develop their English language skills before they begin at Jane's. Instructors recommend that students whose first language is not English attend the workshop series by speaking to students one-on-one or emailing students who they believe could benefit from the support. Some students feel defeated or self-conscious about being encouraged to attend the workshops, so their instructors and I emphasize that the program offers these workshops to support students' success at Jane's while reminding them that attending is their choice. Therefore, by making the workshops optional, students take responsibility for identifying their need for English language support and attending and participating in the workshops.

Courage

While reflecting on the value of courage, I was struck by the courage that students consistently demonstrated in the workshops. Active participation in the workshops requires significant courage as students practice the target language in front of their facilitator and peers. Students demonstrate courage by being willing to take risks and risk failure (ICAI, 2021). Students do not always walk into learning environments feeling courageous. Due to the previously discussed challenges and barriers faced by English language learners in post-secondary institutions, in my experience, students need to be encouraged and supported to feel safe enough to make mistakes and learn from those mistakes. For students to take the required risks to increase their language skills, they need to have a connection with their classmates and their teacher (Government of Manitoba, 2023). In the English language workshops, I aim to establish a safe learning environment so that students have the courage to try without fear of failure. We continue to practice until students are confident with the target language and the associated cultural nuances.

Conclusion

Growing a culture of academic integrity in a polytechnic can encompass change at an institutional level; however, reflecting on our individual contributions is an exercise everyone can participate in at any time. In reflecting on how the English language workshops prepared HTM students for lunch service at Jane's, I have identified the importance of the fundamental values of academic integrity in my teaching practice. By applying the values of honesty, trust, fairness, respect, responsibility, and courage, I have created a workshop culture where students can build the language skills they need to succeed in a fine-dining setting.

Thinking about our impact on institutional culture is valuable, even for staff who may not have a direct appointment related to academic integrity. My understanding of how to practice these values has grown through reflection and discussion with members of the post-secondary education community. I believe academic support staff hold a unique position that allows students to learn about academic integrity in a safe and supportive environment. Therefore, support staff should seek opportunities to incorporate and demonstrate the fundamental values of academic integrity. By demonstrating these values, we set an example for students to follow, encouraging a culture of integrity that supports student success in academic and professional pursuits.

References

Ambrose, S. A., Bridges, M. W., DiPietro, M., Lovett, M. C., & Norman, M. K. (2010). *How learning works: 7 research-based principles for smart teaching*. John Wiley & Sons, Inc.

East, J. (2010). Aligning policy and practice: An approach to integrating academic integrity. *Journal of Academic Language and Learning, 3*(1), 31–51.

Government of Manitoba. (2023). Senior years English as an additional language literacy courses: EAL literacy pilot. https://www.edu.gov.mb.ca/k12/cur/eal/eal-literacy/sy/full_doc.pdf

International Center for Academic Integrity (ICAI). (2021). *The fundamental values of academic integrity*. https://academicintegrity.org/images/pdfs/20019_ICAI-Fundamental-Values_R12.pdf

Kubota, R., Corella, M., Lim, K., & Sah, P. K. (2023). "Your English is so good": Linguistic experiences of racialized students and instructors of a Canadian university. *Ethnicities, 23*(5), 758–778. https://doi.org/10.1177/14687968211055808

Morris, E. J. (2018). Academic integrity matters: Five considerations for addressing contract cheating. *International Journal for Educational Integrity, 14*(15). https://doi.org/10.1007/s40979-018-0038-5

RRC Polytech. (n.d.). *Jane's*. https://www.rrc.ca/janes/

Emilie Jackson is an Academic Support Specialist with the Academic Success Centre at Red River College Polytechnic in Winnipeg, Manitoba, Canada. After completing a BA Honours degree and ESLTCP at the University of Winnipeg, she began working as an English Language Instructor at RRC Polytech's Language Training Centre (now Centre for Newcomer Integration), where she enjoyed teaching international and newcomer students academic language and college-readiness skills to help prepare them for their college program. In January 2021, she joined the Academic Success Centre, where she engages with students in college programs through tutoring, workshops, and resource development. Her primary focus in this role is to support students in strengthening their verbal and written communication skills while developing strategies to help them thrive academically and beyond.

Chapter 17
A Duty of Care in Automotive Technology Education

Christine Gray and Lisa Vogt

Abstract The study of automotive technology is one of the original programs at Red River College Polytechnic in Winnipeg, Manitoba, Canada, dating back to at least 1948. Over the years, the programs, student profiles, automotive industry, and automotive technologies have evolved significantly. Today, the automotive repair industry employs a diverse workforce and commonly promotes values of honesty and commitment to community, making academic integrity more important than ever. In this vocational vignette, a Red Seal Automotive Service Technician shares her experience enacting a duty of care as an educational assistant in Automotive Technologies programs. The strategies discussed include personal readiness, creating healthy learning environments, team and material preparedness, and professional development. The authors illustrate how these strategies have been used to overcome challenges and cultivate a culture of integrity by building a trusting environment where empathy and compassion are prioritized. The chapter concludes with a description of two workshops designed to link academic integrity to industry values, helping students understand the relevance to their future careers.

Keywords Duty of care · Academic integrity · Vocational education · Automotive technology · Automotive technician · International students · Industry safety · Integrity culture · Canada

C. Gray
Red River College Polytechnic, Winnipeg, MB, Canada
e-mail: cgray52@rrc.ca; lvogt@rrc.ca

L. Vogt (✉)
Red River College Polytechnic, Winnipeg, MB, Canada
e-mail: lvogt@rrc.ca

L. Vogt, B. M. Stoesz (eds.), *Academic Integrity in Vocational and Polytechnic Education*, Ethics and Integrity in Educational Contexts 8,
https://doi.org/10.1007/978-3-031-90091-4_17

Introduction

Think about the last time you had your vehicle serviced. Did you receive the necessary information to make informed decisions about maintenance and repairs, or was it simply a transactional exchange where you paid a fee for a service? Which approach would make you more likely to return? Integrity in business hinges on honesty and fairness, and meaningful interactions can significantly enhance business relationships. Just as business integrity can result in successful business ventures and client relationships, meaningful interactions between students and educators can contribute to trusting and respectful educational environments, which promote academic integrity.

In this chapter, Christine Gray and Lisa Vogt collaborate in the writing process to share Christine's experiences as an educational assistant in the Automotive Technologies programs at a polytechnic institution in Winnipeg, Manitoba, Canada. The narrative, told from Christine's first-person perspective, situates academic integrity as a component and a bellwether of student well-being. This chapter uses the concept of a *duty of care* to situate academic integrity as a collective concern, highlighting the responsibility that instructors and institutions have to provide reasonable care for students and avoid harm (Gedajlovic & Wielemaker, 2020; Stoesz, 2024). Christine describes her team's dedication to fostering healthy learning environments that uphold academic integrity, illustrating how a duty of care can be integrated effectively into disciplines not traditionally associated with caregiving and generate benefits for students, the institution, and industry partners.

Christine Gray's Positionality

As an automotive technician and an educational assistant, I support the Transportation Technologies Department staff and students in achieving educational goals. I am part of the female minority in the automotive trades and an immigrant to Canada, and I firmly believe that these elements help me build rapport and trust with students. My role provides academic and personal support, although when I started, I did not anticipate that providing care would be such a significant component of my job. Encouraging and nurturing academic integrity is an aspect of my role that I take seriously. I use a compassionate approach that also upholds our institution's academic integrity policy and prepares students for their future careers in the automotive industry. I believe that future industry professionals can learn to improve their service delivery through conversations about integrity while in their educational program, and educators and support staff play a significant role by demonstrating empathy, compassion, and care.

Our Students

Students in our program may come from a local community or a faraway country, and some are highly interested in working on vehicles, whereas others may have never looked underneath the hood of a car. Some students are very interested in a future career in the automotive trade, but others view the program as a stepping-stone to achieving different goals. Male-identifying students are the majority, but the number of female-identifying students in automotive programs is growing slowly and steadily. From my experience, it is imperative for students to have people who are available to support them, whether they are struggling with their mental health, finances, or academics or are afraid to speak to their instructors. Students have shared with me that they need a place to feel safe, supported, and encouraged as they go through their educational journey and make decisions for their future.

A Duty of Care Applied

At the institutional level, a duty of care results in a range of academic and personal support services that reduce the risk of student failure or harm. At the individual level, I also consider my duty of care and have intentionally prioritized strategies that focus on well-being. Working from this perspective, I have repeatedly observed over multiple years with different cohorts that our instructional team experiences many benefits and that students are able to thrive, complete their studies successfully, and find employment post-graduation. In this section, I outline the elements that I believe are essential for individuals working in higher education to fulfill their duty of care.

Personal Readiness

To carry out my role effectively, I start by considering my own personal readiness to support students. I need the right mental attitude to turn every learning experience, whether good or bad, into success. A mental attitude is "a complex mental state involving beliefs and feelings and values and dispositions to act in certain ways" (Vocabulary.com, n.d.). To be mentally ready, I need to engage in self-care outside the classroom—mentally, physically, spiritually, and relationally. Through trials and tribulations, I have formed a growth mindset where I believe I can continually improve, and this attitude can spill over to those around me.

Healthy Learning Environments

There are many variables that I may not have control over, however, I believe that it is my responsibility to make every effort to create the healthiest learning environment for our students during their time with us, and that starts by demonstrating and promoting integrity every day, one task at a time. There are many reasons we need to create a healthy learning environment that can support all our students' needs, curiosity, and development throughout their journey with us. According to the University of Calgary Taylor Institute for Teaching and Learning (n.d.), "promoting wellbeing and health in your classroom can improve student learning experiences, quality of communication, and overall attainment of course learning outcomes." A healthy learning environment is also important to fostering academic integrity (Bertram Gallant & Kalichman, 2010). With a foundation of integrity, our students can build trust with their employers after graduation, resulting in businesses building trust with their customers. This foundation of integrity is crucial and impacts every person driving a vehicle. If a job is not completed with integrity and quality in our trade, even if it is the smallest task, the customer may need to return their vehicle to the automotive shop, or an accident or serious harm may result. We must make integrity a focus within the automotive industry, and that starts with each of us.

Team Preparedness

To foster a culture of academic integrity, I believe that the more prepared the instructional team is, the easier it will be to achieve academic integrity goals. Preparedness means planning for the upcoming academic year and individual terms and then looking at specific weekly and daily needs. The instructional team must want to be *present* in the classroom and be passionate about building a solid ethical and educational foundation with students. This is the foundation which students build on for the rest of their lives. We also develop a growth mindset by examining past experiences, identifying areas for improvement, and taking action to improve. Our team checks in with each other regularly, finding ways to share responsibilities and strategize together. In these discussions, we determine what we can improve or need to change completely. Our team interactions provide mentorship and professional development, and the environment serves as a support group.

When starting new programs or projects, we sometimes feel like "we are flying by the seat of our pants" due to a lack of time or resources to plan and prepare. At times, we feel like we are crashing and burning. These challenging experiences, however, sharpen skills, build confidence, and bring awareness about different student needs, including academic, personal, and mental well-being. With this learning, our instructional team is better prepared to demonstrate, inspire, and foster a culture of academic integrity. Our growth mentality is fueled by the desire to develop our students into strong, ethical automotive technicians.

Material Preparedness

The instructional team must prepare the shop, classroom materials, and equipment. Material preparedness means having the proper fluids in stock for maintenance and services and rolls of brake lines so that students can practice bending and flaring brake lines for brake line replacement. Due to harsh weather conditions in Manitoba, Canada, rusty part replacements are a large part of a technician's day. In addition, material preparedness includes careful planning of which vehicles will be in the shop so that different instructors can use the same vehicle for different course topics. Thus, communication within the instructional team is essential for creating a positive educational environment.

Professional Development

Knowing you have the support of the instructional team leader is another crucial part of preparing an educational environment that fosters academic integrity. For instructional team members to grow in their roles, regular professional development opportunities are necessary, including growing and maintaining connections with employers to understand changing industry needs. Staying up to date with hiring practices and workplace values in the automotive industry informs how we approach our instructional goals, overall program, and course and lesson plans.

Professional development also includes round table discussions about encouraging academic integrity with other trades programs within the institution and with educators at other institutions, as well as learning new strategies to foster student engagement or support students in distress. We need to know about the types of support services available to our students on our campus to make referrals for academic tutoring, language support, personal counselling, and international student advising. Knowing that the institution has support to care for students beyond our program helps the instructional team do their jobs better. Our preparedness enhances our ability to encourage students to learn with integrity, which will serve them well as they begin their careers as automotive technicians.

A Challenging Situation

Despite our best-laid plans and efforts, some cohorts of students bring more challenges than initially anticipated. In the past, for example, weekday evening classes were offered for cohorts of international students due to daytime class capacity limits. We anticipated that evening programming could pose difficulties not experienced during the day program, as fewer college support services are available, and some staff might find the evening timeslot undesirable. We worked to build community amongst the instructional team and students to reduce the potential difficulties. Unfortunately, community-building efforts do not always reduce the number and types of issues, and at times, an increased number and severity of academic integrity breaches occurred. We experienced situations where students disregarded the instructor's directions, shared their work amongst themselves, and cut corners in the shop. Careful actions are necessary for the safety of the technicians and clients in the shop, such as wearing proper personal protective equipment, following procedures carefully and in the correct order, and working attentively. In addition, when students do not participate fully in classroom and shop activities, they reduce their learning opportunities and decrease their competence in knowledge and skills, which will have negative consequences when students transition to work as technicians in the automotive industry. Our job as an instructional team is to ensure that students can demonstrate safe decision-making, and we expect full participation in learning activities. When students disregard these expectations, the result is an unsafe and challenging teaching and learning environment.

International students who have recently arrived in Canada can be at risk for academic and training difficulties that could lead to program withdrawal (Haverila et al., 2020). International students may be living on their own for the first time, and family support is not in close physical proximity. Many may be under financial stress as they try to cover daily expenses and pay for their studies. Some students are expected to send funds to their families living in their home countries. When financial stresses become a significant burden on students, the instructional team feels helpless as there is little we can do to offer financial support. International students are often not eligible for student loans, and the number of bursaries and scholarships available for international students is limited. The only solution to difficult financial situations is to work, which can lead to time management issues, reduced physical health, anxiety, depression, and fatigue (Centre for Innovation in Campus Mental Health, n.d.). These challenges, in turn, may result in a lack of focus in the classroom and making mistakes that contribute to unsafe learning conditions.

We believe that the cheating and plagiarism we observed within our most difficult cohorts resulted from the students' precarious position as international students with significant financial burdens. As is good practice, international students build communities of support among their classmates to cope with the various challenges and work and learning roles that they are attempting to balance. For most international students, communities are essential and contribute to positive work, study, and life experiences in Canada (Malette & Ismailzai, 2020), however, poor decision-making can occur when the group dynamics are negative. For example, if only one student had time to study and understand the material for a test, they will choose a seat in the classroom that allows their answers to be shared behind them and by as many as possible throughout the classroom. Students might copy practical worksheet answers from each other without completing the tasks themselves or share their full shop activity worksheets, often not handing them in on time. Students who are constantly late or missing school because of work may accuse the instructional team of lacking empathy. These accusations can take an emotional toll on the instructional team; however, we must uphold educational and assessment standards and follow through with consequences for academic integrity breaches as part of our commitment to fostering a culture of academic integrity.

Our Response to a Challenging Situation

The stress of a situation like the one described above was difficult for the instructional team, but we have used these experiences to revisit our teaching, learning, and support approaches and make improvements. We became more flexible with the organization of evening classes because our international students usually worked during the day. For example, we experimented with different start times in the evening so that students could visit academic, personal, and settlement support services before these offices closed at the end of the regular business day. We also gathered answers to our students' questions and shared those answers with our students during their evening classes. Evening cohorts also needed to break for a meal earlier in the evening, and they required longer breaks since on-campus food services were closed, and they had to go off campus if they needed to purchase food. The end of class time also had to be considered carefully, as leaving campus too late in the evening meant some students would miss the last connecting bus to go home.

We also recognized the importance of planning the first week of class to set the tone for the rest of the year. The intention of this first week is to build rapport and a support team that students can access whenever needed. Students and instructors get to know each other during class orientation activities, and we tour the campus together to introduce the campus services and support. During this time, students ask the first questions or make comments out of their own curiosity, which is the first step in building the trusting instructor-student relationships we want to foster. We introduce all our team members to the students, including management, stating that everyone is committed to the students' success. We set expectations for students and encourage them to set their own learning goals.

By advocating for students and making small adjustments, we have been able to anticipate students' needs and improve the classroom culture. When the instructional team advocates on behalf of the student, we enhance trust, showing that instructors' care for students goes beyond the basic requirements to complete assignments, classes, or programs of study.

Academic Integrity Workshops

Our team worked with co-author, Lisa Vogt (academic integrity specialist at that time), to strategize unique ways to capture our students' attention and make academic integrity engaging and relevant. The following two support sessions had a meaningful impact on our students.

In-Class Workshop

The in-class workshop focused on the six fundamental values of academic integrity (ICAI, 2021): honesty, fairness, trust, respect, responsibility, and courage. Students worked in groups to brainstorm the meaning of the six values in our teaching and learning environments (e.g., the classroom and shop). This was an especially helpful activity in a diverse group where individual students can have varying understandings of each value and how to enact them. For example, some students expressed that it is *their responsibility* to make sure their classmates pass an assessment, course, and/or program; however, this belief can lead to breaches of academic integrity. Bringing these ideas forward in the workshop, both instructors and students could express how they interpreted each value, leading to a greater understanding of one another's perspective. Without a common understanding of the fundamental values, students are prone to violating the institutional academic integrity policy, which can lead to long-term negative personal and academic impacts (Davis, 2022).

Academic Integrity Pop-in Session

We also ran a series of short virtual classroom presentations called academic integrity pop-in sessions. Together with the academic integrity specialist, we identified and invited industry partners to a short virtual session. During the first 10 min, the guest from industry spoke about their organization's integrity-related values. Afterwards, the academic integrity specialist spoke for 5 min about the fundamental values of academic integrity, and the session ended with a 5-min question-and-answer period. The real-life scenarios shared by the industry guests made a strong impression on the students, who began to understand that integrity matters outside of the teaching and learning environment. Students commented that these sessions increased their interest in applying to work with the industry guests in the future because the work environment aligned with their personal values. The industry guests were eager to participate as it allowed them to promote their organization to upcoming graduates in hopes of recruiting the best automotive technicians.

Conclusion

We have learned that academic integrity is enhanced when the team actively works to build trust and a strong classroom community. Our commitment to ethical principles resonates throughout the program, and we know that employers actively seek these same values when selecting candidates. The instructional team is responsible for demonstrating and encouraging integrity in the shop, as these are also expected in the industry. Students also show integrity by making efforts to understand the tasks assigned, knowing how and where to find procedures and information, following the procedures safely and properly, and double-checking their work to confirm proper repair and/or maintenance has been done. The steps for proper work in the shop require adherence to the fundamental values of academic integrity — responsibility, honesty, respect, and trust (ICAI, 2021), and placing these values as a foundation during their education and training helps automotive technicians avoid misconduct later during employment. Although our graduates do not take an oath of practice, they complete examinations and a red seal licensing process and hold high value for the safety of vehicles and roadways. As graduates, automotive technicians know that a poorly done job can have serious consequences. With the completion of quality work, rapport and trust are built between technicians and their supervisors. In time, this translates into trust between the employer and their customers, building a good reputation for the business. The combined efforts of good rapport, quality work and trust reduces the number of return visits of unhappy customers, saves money, grows the business, and helps secure the graduate's employment with the company.

I believe that working from a duty of care perspective situates academic integrity as a collective responsibility and highlights the responsibilities that instructors and institutions have towards their students. Fostering a healthy learning environment builds a solid foundation for educators to model standards of high excellence and integrity, coupled with empathy and compassion. When we expect excellence from students, it establishes an environment for students to develop and practice an academic integrity mindset necessary to succeed in the automotive trade. Students follow up with me every year to inform me about their lives and careers and share that our support beyond academics made a difference in their lives, setting them up to have confidence and belief in their abilities to do their jobs with integrity.

References

Bertram Gallant, T., & Kalichman, M. (2010). Academic ethics: A systems approach to understanding misconduct and empowering change in the academy. In T. Bertram Gallant (Ed.), *Creating the ethical academy: A systems approach to understanding misconduct and empowering change*. Routledge.

Centre for Innovation in Campus Mental Health. (n.d.). *Financial vulnerability*. https://campusmentalhealth.ca/toolkits/international-students/financial-vulnerability/

Davis, M. (2022). Examining and improving inclusive practice in institutional academic integrity policies, procedures, teaching and support. *International Journal for Educational Integrity, 18*(1). https://doi.org/10.1007/s40979-022-00108-x

Gedajlovic, E., & Wielemaker, M. (2020). Neither abuse, nor neglect: A duty of care perspective on academic integrity. *Canadian Perspectives on Academic Integrity*, 3(2). https://doi.org/10.11575/cpai.v3i2.71655

Haverila, M., Haverila, K., & McLaughlin, C. (2020). Variables affecting the retention intentions of students in higher education institutions: A comparison between international and domestic students. *Journal of International Students, 10*(2), 358–382. https://doi.org/10.32674/jis.v10i2.1849

International Center for Academic Integrity [ICAI]. (2021). *The fundamental values of academic integrity*. https://academicintegrity.org/images/pdfs/20019_ICAI-Fundamental-Values_R12.pdf

Malette, N., & Ismailzai, E. (2020). Building bridges to better bonds? Bridging and bonding capital development through on-campus club participation among international and domestic students. *Canadian Journal of Higher Education, 50*(4). https://doi.org/10.47678/cjhe.vi0.188817

Stoesz, B. M. (2024). Academic integrity through ethical teaching and assessment: Overview and current trends. In S. E. Eaton (Ed.), *Second handbook of academic integrity* (pp. 1–14). Springer Nature. https://doi.org/10.1007/978-981-287-079-7_89-1

Wright, A. C., & Biswas, S. (n.d.). *Strategies for promoting positive learning environments*. Taylor Institute for Teaching and Learning, University of Calgary. https://taylorinstitute.ucalgary.ca/resources/strategies-for-promoting-positive-learning-environments

Vocabulary.com. (n.d.). Mental attitude. In *Vocabulary.com* dictionary. https://www.vocabulary.com/dictionary/mental%20attitude

Christine Gray is the Work-Integrated Learning (WIL) Coordinator for Transportation programs at Red River College Polytechnic and was previously an Educational Assistant in the Automotive and Transportation programs. Christine grew up in South America and came to Canada without understanding and speaking English. This journey has created a passion for supporting students, especially international students, on their journeys. She is a two-time graduate of Red River College Polytechnic, completing the Automotive Technician Certificate in 2007 and the Automotive Service Technician Red Seal in 2011. She worked as an Automotive Technician for almost a decade before returning to Red River College Polytechnic as an Educational Assistant. Although she always enjoyed the hands-on work environment of the trade, her passion is student support, and she is committed to lifelong learning and development in this field of Student Support. Christine values personal development, resilience, excellence, and integrity and strives to inspire others around her, including faculty and students.

Lisa Vogt has dedicated over 20 years to higher education, beginning her career as an English as an Additional Language (EAL) teacher in Tokyo, Japan, before transitioning to Canadian higher education. Currently, she is an instructor at Red River College Polytechnic, where she teaches leadership and ethics-focused courses. Lisa has published and presented on academic integrity at local, national and international venues, sharing research findings and practitioner insights. The common thread in her work is creating positive and inclusive educational environments that enable students and staff to thrive. She holds a B.A. in Intercultural Studies, a certificate in Teaching English to Speakers of Other Languages (TESOL), and an M.A. in Higher Education Administration and Leadership.

The manufacturer's authorised representative in the EU is Springer Nature Customer Service Centre GmbH, Europaplatz 3, 69115 Heidelberg, Germany. If you have any concerns regarding our products, please contact ProductSafety@springernature.com

Printed and bound by CPI Group (UK) Ltd, Croydon, CR0 4YY
01/10/2026
02225923-0008